Participatory Research and Gender Analysis

New Approaches

Edited by

Nina Lilja, John Dixon, and Deborah Eade

University of Nottingham
Hallward Library

 Routledge
Taylor & Francis Group

LONDON AND NEW YORK

First published 2011
by Routledge
2 Park Square, Milton Park, Abingdon, Oxon, OX14 4RN

Simultaneously published in the USA and Canada
by Routledge
711 Third Avenue, New York, NY 10017

Routledge is an imprint of the Taylor & Francis Group, an informa business

Originally published in August 2008 © Oxfam GB in *Development in Practice.*

First issued in paperback 2013

The preface of this edition © Deborah Eade

This book is a reproduction of *Development in Practice*, vol. 18, issues 4 & 5. The Publisher requests to those authors who may be citing this book to state, also, the bibliographical details of the special issue on which the book was based.

Typeset in Times New Roman by Taylor & Francis Books

All rights reserved. No part of this book may be reprinted or reproduced or utilised in any form or by any electronic, mechanical, or other means, now known or hereafter invented, including photocopying and recording, or in any information storage or retrieval system, without permission in writing from the publishers. 10 07008902.

Participatory Research and Gender Analysis: New Approaches is based on *Development in Practice* Volume 18, Numbers 4 & 5 (August 2008), published by Routledge, Taylor & Francis Group Ltd. We gratefully acknowledge generous financial support for the journal from affiliates of Oxfam International, in particular Oxfam GB. The views expressed in this volume are those of the named individual contributors and not necessarily those of the Series Editor or Publisher.

A summary of each chapter is available in French, Portuguese, and Spanish on the journal's website. To view these and other development resources, please visit: www.developmentinpractice.org

British Library Cataloguing in Publication Data
A catalogue record for this book is available from the British Library

ISBN13: 978-0-415-57768-7 (hbk)
ISBN13: 978-0-415-84911-1 (pbk)

Disclaimer
The publisher accepts responsibility for any inconsistencies that may have arisen in the course of preparing this volume for print.

Contents

CONTENTS

Previous titles from *Development in Practice*
Series Editor: Deborah Eade

Practical Action Publishing, Rugby

**Deconstructing Development Discourse:
Buzzwords and Fuzzwords** (2010)
*Edited by Andrea Cornwall and Deborah
Eade*

Kumarian Press, Bloomfield, CT

**Development and Humanitarianism:
Practical Issues** (2007)
Edited by Deborah Eade and Tony Vaux

**Development and the Private Sector:
Consuming Interests** (2006)
Edited by Deborah Eade and John Sayer

**Development NGOs and Labor Unions:
Terms of Engagement** (2005)
*Edited by Deborah Eade and Alan
Leather*

Oxfam GB, Oxford

**Development, Women, and War: Feminist
Perspectives** (2004)
*Edited by Haleh Afshar and Deborah
Eade*

**Development Methods and Approaches:
Critical Reflections** (2003)
Edited by Deborah Eade

**Development and the Learning
Organisation** (2003) (in association
with the Institute of Development
Studies and Oxfam America)
*Edited by Laura Roper, Jethro Pettit,
and Deborah Eade*

Development and Culture (2002) (in
association with World Faiths
Dialogue on Development)
Edited by Deborah Eade

Development and Cities (2002) (in
association with UNRISD)
*Edited by David Westendorff and
Deborah Eade*

**Development and Agroforestry: Scaling
Up the Impact of Research** (2002) (in
association with the World Forestry
Centre, CGIAR)
*Edited by Steven Franzel, Peter Cooper,
Glenn L. Denning, and Deborah Eade*

Development and Advocacy (2002)
Edited by Deborah Eade

Debating Development: NGOs in the Future (2001) (in association with Oxfam International)*
Edited by Deborah Eade and Ernst Ligteringen

Development, NGOs, and Civil Society (2000)*
Edited by Deborah Eade

Development and Management: Experiences of Value-Based Conflict (2000) (in association with The Open University)
Edited by Deborah Eade, Tom Hewitt, and Hazel Johnson

Development and Social Action (1999)
Edited by Deborah Eade

Development with Women (1999)
Edited by Deborah Eade

Development and Rights (1998)*
Edited by Deborah Eade

Development for Health (1997)
Edited by Deborah Eade

Development and Patronage (1997)*
Edited by Deborah Eade

Development and Social Diversity (1996)*
Edited by Deborah Eade

Development in States of War (1996)*
Edited by Deborah Eade

* Also published in Spanish translation by Intermón, Barcelona.

Contributors

Arega Alene is an agricultural impact economist at the International Institute of Tropical Agriculture (IITA) in Malawi.

Aden A. Aw-Hassan is a senior agricultural economist at the International Center for Agricultural Research in the Dry Areas (ICARDA) in Syria.

Andrew Bartlett has 25 years' development experience in Asia and is currently employed by Helvetas as the Chief Technical Adviser for the Laos Extension for Agriculture Project.

Stephen Biggs is a Research Fellow in the School of Development Studies at the University of East Anglia, and although mainly retired remains professionally active in fair/ethical trade, participatory rural research and development, and science and technology policy and practice.

Mauricio Bellon is Director of the Diversity for Livelihoods Programme at Bioversity International in Italy. He was formerly Human Ecologist at CIMMYT in Mexico.

Robert Delve is a Senior Scientist with the Tropical Soil Biology and Fertility Institute of CIAT, based in Zimbabwe.

John Dixon is Senior Adviser to the Cropping Systems and Economics (CSE) programme and Regional Coordinator for South Asia at the Australian Centre for International Agricultural Research ICIAR). He was previously Director of Impacts Targeting and Assessment at the International Wheat and Maize Improvement Center (CIMMYT) in Mexico.

Morenike Dipelou is a professor in the Department of Veterinary Public Health and Reproduction and Deputy Dean of the College of Veterinary Medicine at the University of Agriculture, Abeoku, Nigeria.

Deborah Eade has 30 years' experience in the international development and humanitarian sector, including 10 years working with various NGOs in Mexico and Central America. She was Editor-in-Chief of *Development in Practice* from 1991 to 2010 and is now working as a freelance writer and editor for international agencies. She has been based in the Geneva area since 1994.

Esbert Friis-Hansen is an economic geographer and senior research fellow at the Danish Institute for International Studies with 25 years' experience as a development researcher and consultant in East and Southern Africa.

CONTRIBUTORS

Delia Grace is a veterinary epidemiologist on joint appointment with the International Livestock Research Institute in Kenya and Cornell University in the USA.

Chanda Gurung Goodrich is the Senior Fellow–Research at the South Asia Consortium for Interdisciplinary Water Resources Studies works on projects related to gender, sustainable livelihoods, and participatory research and development in India and Nepal.

Elisabeth Gotschi works for UNDP in Nicaragua and was previously a research fellow with CIAT and Universität für Bodenkultur (BOKU) in Mozambique.

Jon Hellin works at the International Maize and Wheat Improvement Center (CIMMYT) in Mexico. He has 17 years' research and rural development experience in Latin America, South Asia, and East Africa and previously worked for Practical Action.

Janice Jiggins is a visiting scholar at the Communication and Innovation Studies Group at Wageningen University in The Netherlands.

Scott Justice is an adviser and consultant with the National Agricultural and Environmental Forum of Nepal.

Susan Kaaria is a resource economist and senior scientist at the International Centre for Tropical Agriculture in Uganda.

Erastus Kiambi Kang'ethe is a lecturer in the Department of Public Health, Pharmacology and Toxicology at the University of Nairobi in Kenya.

Nina Lilja is Director of International Agricultural Programs in the College of Agriculture K-State Research and Extension, Kansas State University, and was previously Impact Assessment Economist at the CGIAR Systemwide Program on Participatory Research and Gender Analysis for Technology Development and Institutional Innovation (PRGA Program) in Colombia.

Li Xiaoyun is Professor of Development Studies and Dean of College of Humanities and Development at China Agricultural University.

Noel P. Magor is Head of the International Rice Research Institute (IRRI) Training Centre and visiting fellow in the School of Politics and History at Adelaide University. He was previously PETRRA Project Manager and IRRI Representative in Bangladesh.

Francesca Mancini is a tropical agronomist and works with the FAO on promoting good agricultural practices in Asia. She is also vice-chair of the international Expert Panel on the Social, Environmental and Economic Performance (SEEP) of the world cotton industry, constituted by the International Cotton Advisory Committee (ICAC).

Guy Manners is a science writer and editor with experience in francophone West Africa. He has worked with several of the CGIAR member organisations in Africa and the Middle East.

Mariam Mapila is a regional research fellow and an agricultural economist at the Chitedze Research Station of the International Centre for Tropical Agriculture (CIAT) in Malawi.

Tennyson Magombo is a graduate student at the University of Malawi's Bunda College of Agriculture, working on agro-enterprise development and linking small farmers to markets.

Andreas Neef is a research coordinator and senior scientist with the Thai–Vietnamese–German Collaborative Research Program on 'Sustainable Land Use and Rural Development in Mountainous Regions of Southeast Asia'.

Jemimah Njuki is a senior research fellow at the International Centre for Tropical Agriculture (CIAT) in Zimbabwe.

Janice Olawoye is a professor of rural sociology in the Department of Agricultural Extension and Rural Development, at the University of Ibadan, Nigeria.

Tom Randolph is an agricultural economist and leader of the 'Livestock keeping and human health' project at the International Livestock Research Institute in Kenya.

Joe Remenyi is Visiting Professor at China Agricultural University, having retired from Deakin University and serving until 2007 as the International Director of Research at the International Poverty Reduction Centre in China.

Ganesh Sah is a Senior Scientist agricultural engineer in the Nepal Agricultural Research Council working in research, development, and dissemination of the conservation of resources and agricultural machinery.

Ahmad Salahuddin is a consultant of the International Rice Research Institute (IRRI) in Bangladesh and was previously manager of the PETRRA research programme.

Annita Tipilda is a sociologist at the International Institute of Tropical Agriculture (IITA), where she works on the gender-disaggregated impact of technology adoption on farmers' livelihoods in northern Nigeria.

Victor M. Manyong is an agricultural economist and currently officer in charge at the International Institute of Tropical Agriculture (IITA) in Tanzania, where he has also served on its research for development council.

Paul Van Mele is Program leader for Partnerships, Learning and Innovation Systems at the Africa Rice Center (WARDA), where he works on scaling up farmer-education approaches, including the use of video.

Chapter summaries

Introduction
NINA LILJA AND JOHN DIXON

Participatory research approaches are increasingly popular with scientists working for poverty alleviation, sustainable rural development, and social change. This introduction offers an overview of the special issue of Development in Practice on the theme of 'operationalising participatory research and gender analysis'. The purpose of this special issue is to add value to the discussion of methodological, practical, philosophical, political, and institutional issues involved in using gender-sensitive participatory methods. Drawing on 16 articles, we place some of the main issues, empirical experiences, and debates in participatory research and participatory technology development in the context of implementation, evaluation, and institutionalisation of participatory research and evaluation approaches.

Some common questions about participatory research: a review of the literature
NINA LILJA AND MAURICIO BELLON

This article reviews, through reference to the published literature, some key questions about participatory research. When should participatory research be used? How should participatory research be applied? What about quality of science in participatory research? Are there any institutional issues associated with the use of participatory research? And what are the benefits and costs of participatory research? The article is not a comprehensive literature review on participatory research, it is not meant to set standards for participatory research, nor to define what constitutes 'good' participatory research, but rather it seeks to summarise the realities of implementing participatory research, as discussed and debated by several published authors, and to provide some useful background for this special issue.

The lost 1990s? Personal reflections on a history of participatory technology development
STEPHEN BIGGS

This article traces a history of agricultural participatory research, largely from the author's personal experience. Participatory research in the 1970s was mostly led by disciplinary scientists, and characterised by innovative activities and open academic debate, with some recognition that policy and development practice was a political process. The 1980s saw a shift to learning from past experience, and a participatory mainstream developed, seeking methods for scaling up. Meanwhile, others sought to understand and influence policy and institutional change in their political and cultural contexts, and to keep open the academic debates. The author considers the 1990s as 'lost years', during which mainstream participatory practitioners became inward-looking development generalists, not so interested in learning from others outside their paradigm. The late 2000s provide a chance to re-recognise the political and cultural embeddedness of science and technology; re-introduce strong, widely based disciplines; and learn from past activities that resulted in positive development outcomes (planned or unplanned).

Impact assessment of farmer institutional development and agricultural change: Soroti district, Uganda
ESBERN FRIIS-HANSEN

This article is based on participatory development research conducted in Soroti district of Uganda with the aim of assessing the impact of agricultural development among poor farmers. The central argument is that a combination of farmer empowerment and innovation through experiential learning in farmer field school (FFS) groups, changes in the opportunity structure through transformation of local government staff, establishment of new farmer-governed local institutions, and emergence of a private service provider has been successful in reducing rural poverty. Based on an empirical study of successful adaptation and spread of pro-poor technologies, the study assesses the well-being impact of agricultural technology development in Soroti district. The study concludes that market-based spread of pro-poor agricultural technologies requires an institutional setting that combines farmer empowerment with an enabling policy environment.

No more adoption rates! Looking for empowerment in agricultural development programmes
ANDREW BARTLETT

The debate on empowerment encompasses an older discourse about the intrinsic value of empowerment, and a newer discourse about the instrumental benefits of empowerment; the concept of agency is useful in understanding this distinction. In agricultural development, empowerment efforts are often instrumentalist, viewed as an advanced form of participation that will improve project effectiveness, with adoption rates that promote compliance rather than intrinsic empowerment. Nevertheless, it is possible for projects to enhance the means for – and facilitate the process of – intrinsic empowerment. With regard to process, research and extension can make use of a constructivist rather than the behaviourist approach to support changes in knowledge, behaviour, and social relationships. In assessing empowerment, both developers and 'developees' need to look for evidence that people are taking control of their lives. Case studies – such as those used by the Indonesian Integrated Pest Management (IPM) Programme – will help to capture context and chronology, with unplanned behaviours being particularly useful indicators.

Appraisal of methods to evaluate farmer field schools
FRANCESCA MANCINI AND JANICE JIGGINS

The need to increase agricultural sustainability has induced the government of India to promote the adoption of integrated pest management (IPM). An evaluation of cotton-based conventional and IPM farming systems was conducted in India (2002–2004). The farmers managing the IPM farms had participated in discovery-based ecological training, namely Farmer Field Schools (FFS). The evaluation included five impact areas: (1) the ecological footprint and (2) occupational hazard of cotton production; and the effects of IPM adoption on (3) labour allocation; (4) management practices; and (5) livelihoods. The analysis showed that a mix of approaches increased the depth and the relevance of the findings. Participatory and conventional methods were complementary. The study also revealed different impacts on the livelihoods of women and men, and wealthy and poor farmers, and demonstrated that the value of the experience can be captured also in terms of the farmers' own frames of reference. The evaluation process consumed considerable resources, indicating that proper budgetary allocations need to be made.

Engaging with cultural practices in ways that benefit women in northern Nigeria
ANNITA TIPILDA, AREGA ALENE, AND VICTOR M. MANYONG

This study explores the intra-household impact of improved dual-purpose cowpea (IDPC) from a gender perspective, in terms of productivity and food, fodder, and income availability, the impact of which is linked to the income thus placed in the women's hands. Surplus income is important in providing food and nutritional benefits to the home, particularly during periods of risk. More importantly, income generated through the adoption of improved cowpea varieties has entered a largely female domain, where transfers of income reserves were passed on between women of different ages, with significant impact in terms of social and economic development. However, the technology has strengthened the separation of working spheres between men and women. Future technologies should, from the outset, explore provisions existing within the local rubric, to focus on women with the aim of expanding their participation in agriculture with the associated benefits to their families.

Strategies for out-scaling participatory research approaches for sustaining agricultural research impacts
ADEN A. AW-HASSAN

The popularity of participatory research approaches is largely driven by the expected benefits from bridging the gap between formal agricultural science institutions and local farm communities, making agricultural research more relevant and effective. There is, however, no certainty that this approach, which has been mainly project-based, will succeed in transforming agricultural research in developing countries towards more client-responsive, impact-oriented institutions. Research managers must consider appropriate strategies for such an institutional transformation, including: (1) careful planning of social processes and interactions among different players, and documenting how that might have brought about success or failure; (2) clear objectives, which influence the participation methods used; (3) clear impact pathway and impact hypotheses at the outset, specifying expected outputs, outcomes, impacts, and beneficiaries; (4) willingness to adopt institutional learning, where existing culture and practices can be changed; and (5) long-term funding commitment to sustain the learning and change process.

Integrating participatory elements into conventional research projects: measuring the costs and benefits
ANDREAS NEEF

Until recently, participatory and conventional approaches to agricultural research have been regarded as more or less antagonistic. This article presents evidence from three sub-projects of a Thai–Vietnamese–German collaborative research programme on 'Sustainable Land Use and Rural Development in Mountainous Regions of Southeast Asia', in which participatory elements were successfully integrated into conventional agricultural research as add-on activities. In all three sub-projects the costs of studying local knowledge or enhancing farmers' experimentation consisted of additional local personnel, opportunity costs of participating farmers' time, and travel costs. However, these participatory elements of the research projects constituted only a small fraction of the total costs. It may be concluded that conventional agricultural research can be complemented by participatory components in a cost-effective way, while producing meaningful benefits in terms of creating synergies by blending scientific and local knowledge, scaling up micro-level data, and highlighting farmers' constraints affecting technology adoption.

Participatory research practice at the International Maize and Wheat Improvement Center (CIMMYT)
NINA LILJA AND MAURICIO BELLON

This study assessed the extent to which participatory methods had been used by CIMMYT, and how the scientists perceived them. Results suggest that participatory approaches at the Center were largely 'functional' – that is, aimed at improving the efficiency and relevance of research – and had in fact added value to the research efforts. The majority of projects surveyed also placed emphasis on building farmers' awareness. This is understandable if we think that the limiting factor in scientist–farmer exchange is the farmers' limited knowledge base. Thus, in situations such as marginal areas and in smallholder farming, exposure to new genotypes and best-bet management options would be a first requirement for effective interactions and implementation of participatory approaches.

Making poverty mapping and monitoring participatory
LI XIAOYUN AND JOE REMENYI

The real experts on poverty are poor people, yet the incidence and trends in poverty are usually measured by the use of official economic indicators assumed by researchers to be relevant. Poor householders themselves distinguish between subsistence and cash income. In a 'self-assessed poverty' exercise, poor villagers in rural China specified and weighted key poverty indicators. Eight key indicators describing three basic types of poverty were isolated and used to construct a participatory poverty index (PPI), the components of which provide insights into core causes of poverty. Moreover, the PPI allows direct comparison of the incidence of poverty between villages – differences in social, cultural, and environmental characteristics of each village notwithstanding. As a result, the PPI offers an objective method of conducting poverty monitoring independently of physical and social features. This article provides a brief description of the PPI and the data needed to construct a village-specific PPI.

Participatory risk assessment: a new approach for safer food in vulnerable African communities

DELIA GRACE, TOM RANDOLPH, JANICE OLAWOYE, MORENIKE DIPELOU, AND ERASTUS KANG'ETHE

Women play the major role in food supply in developing countries, but too often their ability to feed their families properly is compromised; the result is high levels of food-borne disease and consequent limited access to higher-value markets. We argue that risk-based approaches – current best practice for managing food safety in developed countries – require adaptation to the difficult context of informal markets. We suggest participatory research and gender analysis as boundary-spanning mechanisms, bringing communities and food-safety implementers together to analyse food-safety problems and develop workable solutions. Examples show how these methodologies can contribute to operationalising risk-based approaches in urban settings and to the development of a new approach to assessing and managing food safety in poor countries, which we call 'participatory risk analysis'.

Pro-poor values in agricultural research management: PETRRA experiences in practice

AHMAD SALAHUDDIN, PAUL VAN MELE, AND NOEL P. MAGOR

PETRRA was an agricultural research-management project which used a values-based approach in project design, planning, and implementation. Through an experiential learning process, agricultural research and development (R&D) institutes, NGOs, private agencies, and community-based organisations rediscovered and improved the understanding of their strengths in meeting development commitments. The project successfully showed how valuesbased research can meaningfully be implemented and a sustainable pro-poor impact achieved.

Operationalising participatory research and farmer-to-farmer extension: the Kamayoq in Peru

JON HELLIN AND JOHN DIXON

While rural poverty is endemic in the Andean region, structural adjustment programmes have led to a dismemberment of agricultural research and extension services so that they are unable to serve the needs of smallholder farmers. The NGO Practical Action has been working in the Andes to address farmers' veterinary and agriculture needs. The work has included the training of farmer-to-farmer extension agents, known locally as Kamayoq. The Kamayoq have encouraged farmer participatory research, and local farmers pay them for their veterinary and crop advisory services in cash or in kind. The Kamayoq model is largely an unsubsidised approach to the provision of appropriate technical services and encouragement of farmer participation. The model also illustrates that, in the context of encouraging farmer participation and innovation, NGOs have advantages over research organisations because of their long-term presence, ability to establish trust with local farmers, and their emphasis on social and community processes.

Using community indicators for evaluating research and development programmes: experiences from Malawi

JEMIMAH NJUKI, MARIAM MAPILA, SUSAN KAARIA, AND TENNYSON MAGOMBO

Evaluations involving stakeholders include collaborative evaluation, participatory evaluation, development evaluation, and empowerment evaluation – distinguished by the degree and depth of involvement of local stakeholders or programme participants in the evaluation process. In community participatory monitoring and evaluation (PM&E), communities agree programme objectives and develop local indicators for tracking and evaluating change. PM&E is not without limitations, one being that community indicators are highly specific and localised, which limits wide application of common community indicators for evaluating programmes that span social and geographic space. We developed community indicators with six farming communities in Malawi to evaluate a community development project. To apply the indicators across the six communities, we aggregated them and used a Likert scale and scores to assess communities' perceptions of the extent to which the project had achieved its objectives. We analysed the data using a comparison of means to compare indicators across communities and by gender.

Participatory technology development in agricultural mechanisation in Nepal: how it happened and lessons learned

CHANDA GURUNG GOODRICH, SCOTT JUSTICE, STEPHEN BIGGS, AND GANESH SAH

International Wheat and Maize Improvement Center (CIMMYT) projects on new resource-conservation technologies (RCTs) in the Indo-Gangetic Plains of Nepal aimed to strengthen equity of access, poverty reduction, and gender orientation in current rural mechanisation processes – more specifically, to promote machine-based resource conservation and drudgery-reduction technologies among smallholder farmers. These projects, together with other projects and other actors, gave rise to an informal 'coalition' project, which used participatory technology development (PTD) approaches, where farmers, engineers, scientists, and other partners worked towards equitable access to new RCTs. This experience showed that PTD projects need to be flexible, making use of learning and change approaches. Once successful adoption is occurring, then what? Such projects need to ensure that everyone is benefiting in terms of social inclusion and equity; this might necessitate new unforeseen work.

Gender equity and social capital in smallholder farmer groups in central Mozambique

ELISABETH GOTSCHI, JEMIMAH NJUKI, AND ROBERT DELVE

This case study from Bu'zi district, Mozambique investigated whether gender equality, in terms of male and female participation in groups, leads to gender equity in sharing of benefits from the social capital created through the group. Exploring the complex connection between gender, groups, and social capital, we found that gender equity is not necessarily achieved by guaranteeing men and women equal rights through established by-laws, or dealing with groups as a collective entity. While there were no significant differences in the investment patterns of men and women in terms of participation in group activities and contribution of communal work, access to leadership positions and benefits from social capital were unequally distributed. Compared with men, women further found it difficult to transform social relations into improved access to information, access to markets, or help in case of need.

Preface

Deborah Eade

For development and humanitarian aid agencies that are committed to gender equity, it is critical to ensure that those who are most affected by – and particularly those intended to benefit from – their interventions can participate in shaping them. Creating opportunities to achieve such participation is all the more important given that these agencies are unelected, and may even be unknown to the people they aim to serve. Partnership in aid should, then, imply mutual accountability and shared responsibility for the outcomes (Eyben 2006).[1] Evidently this does not mean consulting local 'stakeholders' about decisions that have already been taken, as this would amount to little more than co-option into a predetermined agenda. Rather, a gender-sensitive analytical framework ought to translate into ways of working that are based on social inclusion and are therefore open to adaptation in response to insights and concerns that might otherwise have been overlooked, ignored, or misinterpreted – however inconvenient such changes might be for the logframe or 'project cycle.'

Of course the reality is seldom a smooth process. What level and forms of participation are appropriate, and who decides when 'enough is enough'? What methods will provide the best opportunity for meaningful participation in any given circumstances? Whose views will prevail if there is no consensus? What if some people don't show any interest in active participation? And what if these non-participants are mainly women, or from a scheduled caste or ethnic minority, are elderly, or are stigmatised because of what they do for a living, or are (undocumented) migrants whose survival depends on *not* getting involved?[2] And what if, despite all best efforts, the intervention doesn't make any difference to prevailing power structures? Or is undermined by incompetence or prejudice on the part of the agency staff? Where do the lines of accountability run in the event of failure to achieve the intended outcomes, or even leaving things worse than they were before?

Though most development agencies would regard participation as a good thing, at least as a means if not as an end in itself, the *practice* of participation has come under considerable fire, not least among contributors to *Development in Practice*. A tiny sample of the 345 search results on articles the journal has published on the subject over the last 20 years include Anacleti 1993; White 1996, Jackson 1997; Ngunjiri 1998; Mompati and Prinsen 2000; Tate 2004; Simon *et al.* 2005; Leal 2007; and Kamruzzaman 2009 (see also the classic critique by Cooke and Kothari 2001 and the counterarguments by Hickey and Mohan 2005). Similarly, and especially with the widespread adoption of a Rights-Based Approach to development, while any serious aid agency by now has some form of gender policy, arguments about how best to achieve equality between women and men continue – and indeed there debates about whether the goal is gender equality or gender equity[3] and what, if any, relationship the development industry has with any of the many forms of feminism (Smyth 2007; Cornwall *et al.* 2007). Almost the length of a professional career after the 1975 First World Conference on Women in Mexico City, it is still being asked whether we should really be talking about Women *in* Development, Women *and* Development, Gender *and* Development, or Human Development more broadly – and whether gender equality is more likely to be achieved by 'mainstreaming' versus maintaining a battery of 'gender experts', or whether both are necessary; whether 'femocrats' and women's movements are mutually supportive, antagonistic, or simply occupy different realms; whether women's empowerment is a more likely outcome if aid agencies work with men, focus only on women, work with same-sex groups, or with mixed groups of women and men; and whether, given that they can only mirror their own societies, aid agencies themselves need to have affirmative-action policies, quota systems, etc. so that they unambiguously embody their professed commitment to gender equality. Leaving aside Southern feminist critiques of mainstream development per se, most cogently argued by DAWN[4] (and echoed by post-development thinkers), despite 35 years of concerted efforts by governments, aid agencies, women's and feminist movements, and NGOs, an enormous body of scholarship and empirical research, many impressive achievements notwithstanding, discrimination against girls and women persists the world over.[5]

This *Development in Practice Book* addresses the iterative linkages between gender analysis and participatory research with a particular focus on agricultural systems. It is based on the August 2008 double issue of the international journal *Development in Practice* that was guest-edited by Nina Lilja, then Impact Assessment Economist at the CGIAR Systemwide Program on Participatory Research and Gender Analysis for Technology Development and Institutional Innovation (PRGA Program) in Colombia and John Dixon, then Director of Impacts Targeting and Assessment at the International Wheat and Maize Improvement Center (CIMMYT) in Mexico.[6]

Their comprehensive introductory chapter sets out the rationale for what they call 'the broad portfolio of approaches, practices, and frameworks' covered by the 16 chapters, which offer 'an operational context for better-quality implementation, evaluation, and institutionalisation of participatory approaches for the development and research community'. The chapters include a literature review, an annotated compilation of further resources and a blend of theoretical and overview papers and empirical studies illustrating experience from a broad range of countries and regions: Bangladesh, China, India, Nepal, and Indonesia; Malawi, Mozambique, Nigeria, and Uganda; Mexico and Peru.

This is the first in the Routledge series *Development in Practice Books*, which will be based on special themed issues of the journal. This replaces the earlier *Development in Practice Readers* series, which was launched in 1996 as a series of thematic anthologies based on articles drawn from the journal, supplemented by an annotated selection of further resources on the chosen topic, combining original and reprinted material in a stand-alone form. Since then, 21 *Readers* have been published on issues ranging from human rights to civil society, from labour unions to learning organisations, from sustainable urban development to women and war. Many of these have been used for university teaching and in training workshops, and six have also been published in Spanish translation. A full listing is given on page vi, and several are available in full-text PDFs at www.developmentinpractice.org. The new *Development in Practice Books* series will be based solely on themed issues of *Development in Practice* and will therefore include annotated resources only if these formed part of the original issue. With the wealth of resources now freely available via the Internet, we believe that individual readers are in the best position to develop their own lists of further resources in the language(s) of their choice.

Notes

1 Unfortunately, this is not general practice among official donors or NGOs (see for example Mawdsley *et al.* 2005; Wallace *et al.* 2007).
2 In the special issue of *Development in Practice* on Active Citizenship, guest editor Matthew Clarke (2009) discusses this in relation to Burmese migrant labourers in Thailand. Their need to 'lie low' poses real operational challenges for an NGO that has been working with such populations for many years. It clearly challenges assumptions about participation. See also Mompati and Prinsen (2000) for an example of self-exclusion by virtue of 'inappropriate' participation.
3 Some people hold that gender equality refers to outcomes (absence of discrimination, whether overt or implicit) while gender equity describes opportunities or inputs. In practice the two terms are used interchangeably.
4 DAWN (Development Alternatives with Women for a New Era) is a network of women scholars and activists from the economic South who engage in feminist research and analysis of the global environment and are committed to working for economic justice, gender justice, and democracy.
5 For example, the UK Women and Work Commission reported in July 2009 that there has been no significant decline in gender stereotyping among children under the age of 14, and that the gap between what men and women are paid is now increasing after a period when it appeared to be closing slightly: women's median hourly earnings are currently 22.6% less than men's (Sparrow 2009). Lower incomes throughout their working lives can all too easily translate into economic hardship when women reach pensionable age.
6 Nina Lilja is now Director of International Agricultural Programs in the College of Agriculture K-State Research and Extension, Kansas State University. John Dixon is now Senior Adviser to the Cropping Systems and Economics (CSE) programme and Regional Coordinator for South Asia at the Australian Centre for International Agricultural Research ICIAR).

References

Anacleti, Odhiambo (1993) 'Research into local culture: implications for participatory development', *Development in Practice* 3(1): 44–48.

Clarke, Matthew (2009) 'Over the border and under the radar: can illegal migrants be active citizens?', *Development in Practice* 19(8): 1064–1078.

Cornwall, Andrea, Elizabeth Harrison, and Ann Whitehead (eds.) (2007) *Feminisms in Development: Contradictions, Contestations and Challenges*, London: Zed Books.

Craig, David and Doug Porter (1997) 'Framing participation: development projects, professionals, and organisations', *Development in Practice* 7(3): 229–236.

Cooke, Bill and Uma Kothari (eds.) (2001) *Participation: The New Tyranny?*, London: Zed Books.

Eyben, Rosalind (ed.) (2006) *Relationships for Aid*, London: Earthscan.

Hickey, Samuel and Giles Mohan (eds.) (2005) *Participation – From Tyranny to Transformation? Exploring New Approaches to Participation in Development*, London: Zed Books.

Howard, Patricia L. (2002) 'Beyond the "grim resisters": towards effective gender mainstreaming through stakeholder participation', *Development in Practice* 12(2):164–176.

Jackson, Cecile (1997) 'Sustainable development at the sharp end: field-worker agency in a participatory project', *Development in Practice* 7(3): 237–247.

Kamruzzaman, Palash (2009) 'Poverty Reduction Strategy Papers and the rhetoric of participation', *Development in Practice* 19(1): 61–71.

Kapoor, Dip (2005) 'NGO partnerships and the taming of the grassroots in India', *Development in Practice* 15(2): 210–215.

Leal, Pablo Alejandro (2007) 'Participation: the ascendancy of a buzzword in the neo-liberal era', *Development in Practice* 17(4&5): 539–548.

Mawdsley, Emma, Janet G. Townsend, and Gina Porter (2005) 'Trust, accountability, and face-to-face interaction in North–South NGO relations', *Development in Practice* 15(1): 77–82.

Mompati, Tlamelo and Gerard Prinsen (2000) 'Ethnicity and participatory methods in Botswana: some participants are to be seen and not heard', *Development in Practice* 10(5): 625–637.

Ngunjiri, Eliud (1998) 'Participatory methodologies: double-edged swords', *Development in Practice* 8(4): 466–470.

Pickard, Miguel (2007) 'Reflections on relationships: the nature of *partnership* according to five NGOs in southern Mexico', *Development in Practice* 17(4&5): 575–581.

Simon, David, Duncan F.M. McGregor, Kwasi Nsiah-Gyabaah, and Donald A. Thompson (2003) 'Poverty elimination, North–South research collaboration, and the politics of participatory development', *Development in Practice* 13(1): 40–56.

Smyth, Ines (2007) 'Talking of gender: words and meanings in development organisations', *Development in Practice* 17(4&5): 582–588.

Sparrow, Andrew (2009) 'Mind the gap: UK gender pay divide widens', *Guardian*, available at: http://www.guardian.co.uk/society/2009/jul/29/gender-pay-divide-women-inequality (retrieved 11 August 2009).

Tate, Janice (2004) 'Participation and empowerment: reflections on experience with indigenous communities in Amazonas, Brazil', *Development in Practice* 14(3):407–412.

Wallace, Tina, Lisa Borstein, and Jennifer Chapman (2007) *The Aid Chain: Coercion and Commitment in Development NGOs*, Rugby: Practical Action Publishing.

White, Sarah C. (1996) 'Depoliticising development: the uses and abuses of participation', *Development in Practice* 6(1): 6–15.

Introduction

Nina Lilja and John Dixon

Participatory research approaches are increasingly popular with scientists working for poverty alleviation, sustainable rural development, and social change. This introduction offers an overview of the special issue of Development in Practice *on the theme of 'operationalising participatory research and gender analysis'. The purpose of this special issue is to add value to the discussion of methodological, practical, philosophical, political, and institutional issues involved in using gender-sensitive participatory methods. Drawing on 16 articles, we place some of the main issues, empirical experiences, and debates in participatory research and participatory technology development in the context of implementation, evaluation, and institutionalisation of participatory research and evaluation approaches.*

The function of development research is to generate new knowledge or apply existing knowledge in new ways that can be used – in the context of the development process – to increase people's welfare and, in doing so, to eradicate poverty. It is widely acknowledged that successful research on agriculture and natural-resource management requires dialogue and co-operation between those who produce knowledge (technology) and the decision makers (end-users, farmers) who use it. This is the rationale for the use of participatory methods and gender analysis in research and development efforts targeting poverty alleviation, social inclusion, and equity. Participatory development arose as a reaction to the failure to involve would-be beneficiaries of 'development' in the process. As such, it was popularised by, for example, Chambers *et al.* (1989), Chambers and Conway (1992), and Chambers (1994, 1995).

Underpinning the rise of participatory research has been a realisation that the poor in general, and poor marginal farmers in particular, are far from being a homogeneous group. Thus, technologies have to be selected and adapted for particular systems, ideally with strong engagement if not full control by women and men farmers (see, for example, Ashby and Sperling 1995; Ceccarelli and Grando 2007; de Jager *et al.* 2004; Witcombe *et al.* 1999). Such interaction constitutes a two-way knowledge pathway between researchers and agricultural communities.

1

It therefore follows that the research and development process in which they engage is based on joint learning and production of knowledge. The essence of participatory research is not greater involvement as subjects in conventional research *per se*. It should not be regarded as an alternative methodology: participatory research is an approach that can be applied to a wide range of methodologies – surveys, experiments, impact assessments, monitoring, and evaluation – and is characterised by three defining elements: science, co-learning, and action.

This special issue of *Development in Practice* is concerned with operationalising participatory research and gender analysis. Recently a group of research and development professionals defined three areas of knowledge with regard to improving the understanding of the complexity of poverty and change, which are equally relevant to operationalising gender-sensitive participatory research aimed at poverty alleviation and social inclusion (Kristjanson *et al.* 2008). These three areas are: linking research with action; impact assessment and evaluation; and institutionalisation of new research and assessment and evaluation approaches. *Linking research with action* is concerned with how research reaches end users; it is based on premises that sustainable development will depend in large part on an improved dialogue between the science and technology community and problem-solvers applying technology in pursuit of sustainability (and development) goals (Clark *et al.* 2002: 6). The propositions relate to providing a user-driven dialogue between scientists and decision makers to define the research/development problem and to design interventions, including complex issues such as power relations and multiple partnerships, as well as issues related to political, economic, and cultural context. *Impact assessment and evaluation* propositions deal with assessing the impact of participatory research, using participatory impact-assessment and evaluation methods, as well as issues about attribution and uses of evaluation results. *Institutionalisation* propositions are concerned with the mainstreaming of participatory approaches in research and impact assessment/evaluation; they are concerned with the changes in institutions and behaviours necessary for sustainable development and social inclusion.

Linking research with action: implementation of gender-sensitive participatory research

For more than two decades, the literature has been replete with a variety of approaches and tools for participatory research and gender analysis. Yet the harsh reality is that projects utilising participatory methods struggle with a multitude of challenges, such as failure to choose an appropriate tool for a particular 'research/development problem'; mechanical or poor-quality application of participatory tools; and sometimes a lack of credibility of the approach in the eyes of users and the development community. At the core of this challenge, research and development professionals struggle to expand their understanding of the complexities of the 'research/development problem' and complicated pathways to achieve research/development impact, as well as the design of suitable interventions which will lead to impact. Perhaps not enough attention has been paid in the past to the chronological point at which communities and local people become involved in decision-making processes. The point of their engagement affects the kinds of participation employed and the subsequent impact pathways taken. Problem definition is a key moment, because it is the point at which various participants (community members, researchers, and others) establish basic assumptions about the nature of problem(s), decide on the inclusion and exclusion of local communities and partners, and define potential solutions (interventions) to be pursued.

The two-way knowledge pathway between researchers and agricultural communities is the foundation of participatory research and can lead to various impact pathways. In reality, such knowledge pathways often take the form of networks of rural social actors which operate as innovation systems to adapt and transform technologies, manifested as the enrichment of knowledge and strengthening of flows among the actors and along the researcher–farmer pathways. Such networks can also be viewed as 'messy partnerships' (Guijt 2008) or 'complex' systems (Rogers 2008), which perform a crucial function of spanning the boundaries between research and agricultural production domains (Band and Jax 2007; Dickson 2008). Complex systems comprise components with multiple feedback loops, characterised by conditional and uncertain relationships (Kauffman 1996). As such, there is no simple linear change–effect relationship, and the managers seeking to improve these researcher–farmer pathways are obliged to pursue response–act–observe strategies (Snowden and Boone 2007). The two-way knowledge flows along the researcher–farmer pathway can also be visualised as boundary-spanning objects, or as vertical or horizontal learning platforms (Dixon 2004) which link communities, countries, or regions.

Impact assessment and evaluation

Perhaps the first of the major changes facing impact assessors has been the extension of participatory research methods beyond partnering fellow researchers in the national programmes to include farmers, the ultimate beneficiaries of the research (and development) process. Early participatory methods focused on the *functional* aspect of improving the research process by involving the intended end-users in various stages to ensure that the end-products (technologies, mainly crop varieties and management practices) were best suited to clients' needs and systems. This in turn was expected to result in faster and higher rates of adoption and impact of the new technologies – and in many cases succeeded in doing so. More recently, participatory methods have been used to *empower* the intended beneficiaries, enabling them to assess their own needs and address them either directly or by creating demands on research organisations (Hellin *et al.* 2008).

For participatory research to be effective in advancing poverty alleviation and social inclusion, its accountability must extend beyond donors and peer reviews. The outcomes of participatory processes therefore need to be sufficiently reliable and relevant to people's needs, in order to justify the cost of participation; moreover, they need to be sufficiently democratic to avoid gender, ethnic, and poverty biases. Participatory research is implemented in partnerships, and the process involves collaboration, negotiation, and team building. Impact assessment of such efforts should therefore recognise that *how* the research is carried out is just as important as what is done (Kristjanson *et al.* 2008).

Addressing issues of 'how' change comes about and 'who' benefits involves both quantitative and qualitative methods. Conventional impact-assessment methods are well suited and have been widely used for documenting functional outcomes of participatory research (for example, adoption impacts), but this provides only a partial picture of impact. The use of participatory methods in impact assessment and evaluation addresses a broader notion of impact of the participatory research efforts. New innovative and participatory assessment methods are better suited to assess multiple factors and dynamic interactions, in addition to documenting direct project-intervention outputs and outcomes, and ultimate impact. An increase in documented evidence of the use of such approaches shows a shift from predominantly economic views towards examining multiple factors and their evidence, as informed by broader systems-based impact-assessment and evaluation frameworks.

Institutionalisation of new research and assessment/evaluation approaches

While the technical issues addressed by the research or development intervention are often viewed as the core solution to the 'research/development problem', in the real world the critical elements of the 'problem' is often the institutional context of the research and development intervention at local, meso, and higher levels. The intense focus of participatory approaches on the community level alone diverts attention from political, economic, and cultural realities that constrain the ability of community members and researchers to participate fully, or constrain the available impact pathways for the project to follow.

Much useful knowledge that agricultural communities possess remains untapped, owing to the failure of research and development systems and institutions to encourage and professionally value (and, in many cases, even allow) the use of participatory approaches. Equally, there is much potentially useful knowledge that is laboriously produced through participatory research but never applied, because this knowledge is being produced in isolation from a larger institutional research and development effort. Unless these institutional and individual constraints are relaxed, the contribution of gender-sensitive participatory approaches to poverty alleviation and social inclusion will remain below its potential. Therefore, in the interests of supporting effective operationalisation, methodological innovation, and institutionalisation of participatory research, there is value in further adaptation, documentation, and sharing of participatory approaches and tools in line with the growing diversity of applications and institutional settings.

Background to this issue

In October 2005, the CGIAR Systemwide Program on Participatory Research and Gender Analysis for Technology Development and Institutional Innovation (PRGA Program) and the International Maize and Wheat Improvement Center (CIMMYT) brought together about 30 impact assessors drawn from the research-for-development sector to discuss the *status quo* of agricultural impact assessment of participatory research and gender analysis, and options for its future (Lilja *et al.* 2006). This special issue presents six papers presented at that workshop which discuss opportunities afforded by effective (knowledge) pathways linking researchers and farmers, underpinned by participatory research (and gender analysis). In addition, a set of eight practical notes (selected after an open call) enlarges the portfolio of available approaches and tools presented in this issue for different contexts.

This special issue is organised as follows. Immediately after this introductory essay, we have included a brief literature review by **Lilja and Bellon**, addressing some common questions about the participatory research, followed by a personal reflection of a history of participatory technology development by **Stephen Biggs**. We then present the six workshop papers and the eight practical notes. An annotated resources section concludes the issue. The remainder of this overview provides an introduction to the workshop papers and practical notes, as well as a summary of the salient characteristics of these papers in operationalising participatory research and gender analysis.

Introduction to workshop papers: enriching the operational knowledge base and measuring impacts

In a paper describing participatory research involving an impact assessment of agricultural technology, farmer empowerment, and changes in opportunity structures in Soroti District, Uganda, **Esbern Friis-Hansen** argues that rural poverty has been reduced by a combination of farmer-empowerment and innovation through experiential learning in farmer field school (FFS) groups, and changes in the opportunity structure through transformation of local government staff, establishment of new farmer-governed local institutions, and emergence of private service providers.

Based on an empirical study of successful adaptation and spread of pro-poor technologies, the study assesses the well-being impact of agricultural technology development. It further analyses the socio-economic and institutional context under which pro-poor technologies are adopted by poor farmers. Agricultural growth among poor farmers in Soroti district has been the key factor in poverty alleviation.

The study shows that farmers who are members of FFS groups are significantly better off than non-member farmers. The area-specific well-being ranking methodology (based on farmers' perceptions) proved to be a useful impact-assessment technique. Qualitative interviews further indicate that most farmers were among the 'very poor' or 'poor' when they joined FFS. This is a major achievement, providing evidence in support of the hypothesis that farmer empowerment through demand-driven advisory services can contribute significantly to alleviating rural poverty. The analysis further shows the diversity of pathways out of poverty, including labour, food production, and investment in cattle. A lesson learned is that market-based spread of pro-poor technologies requires an institutional setting that combines farmer and community empowerment with an enabling policy environment.

Andrew Bartlett introduces the popular discussion topic of empowerment and assesses the shift from adoption rates to empowerment with reference to field activities in Laos. The debate on empowerment encompasses two distinct discourses: an older discourse about the intrinsic value of empowerment, and a newer discourse about the instrumental benefits of empowerment. The concept of *agency* is useful in understanding this distinction. In agricultural development, empowerment is often instrumentalist, viewed as an advanced form of participation that will improve project effectiveness. Goals are usually set by the technical and political elite, and adoption rates promote compliance rather than real (i.e. intrinsic) empowerment.

If farmers are to become empowered through their own agency, agricultural development projects can support them in two ways: by enhancing the means and facilitating the process of empowerment. With regard to the latter, research and extension could make use of a constructivist rather than a behaviourist approach to support changes in knowledge, behaviour, and social relationships. Donors and implementing organisations cannot decide the precise outcomes of empowerment for rural people; farmers have to do that for themselves. The use of multi-method, multi-site case studies is examined in some detail, with an example from the FAO programme for integrated pest management (IPM).

Francesca Mancini and Janice Jiggins appraise the methodology of an evaluation study of IPM FFSs in central India. The need to increase agricultural sustainability has induced the government of India to promote the adoption of IPM. An evaluation of cotton-based conventional and IPM farming systems was conducted in India (2002–2004). The farmers managing the IPM farms had participated in a discovery-based learning experience centred on development of ecological understanding, management skills, and group action, namely Farmer Field Schools. The evaluation included five impact areas: (1) the ecological footprint and (2) occupational hazard of cotton production; and the effects of IPM adoption on (3) labour allocation, (4) management practices, and (5) livelihoods. This paper critically reviews the methodologies used in the assessment. The role of participatory research in impact assessment and its synergy with conventional research are discussed. The importance of a gender-sensitive approach to evaluation is analysed on the basis of the experience gained. The analysis shows that a mix of approaches increased the depth and relevance of the findings. The inclusion of women in the evaluation highlighted important gender issues related to the adoption of IPM and the reduction in pesticide use that otherwise would have been overlooked. The evaluation process was resource-consuming, indicating that proper budgetary allocations need to be made if thorough evaluation is to be routinely attempted.

At the heart of feminist participatory research is the fundamental question about the implications of our work for the redistribution or consolidation of power among and between the world's women and men. In a paper on the intra-household impact of improved dual-purpose cowpea (IDPC) on women in northern Nigeria, **Annita Tipilda, Arega Alene, and Victor Manyong** approach the subject from a gender perspective. Increased productivity of IDPC has increased food, fodder, and income availability. The impact of this is linked to the income that it has placed in the women's hands. Surplus income is extremely important in providing food and nutritional benefits to the home, particularly during periods of risk. Most importantly, income generated through the adoption of IDPC has entered a largely female domain, where transfers of gifts and income reserves are invested in assets that are passed between generations, thereby having meaningful impact on the social and economic development of women – although the technology reinforced the separation of men's and women's working spheres. Consequently, future technologies to be developed by the International Institute for Tropical Agriculture (IITA) should attempt from the onset to use existing alternatives, from within the local rubric, to target women, with the aim of expanding their participation and contribution to agriculture with the associated benefits to their families.

Aden Aw-Hassan takes on the perennial challenge of out-scaling (and scaling up) participatory research approaches, with particular reference to drylands, and concludes that learning is a crucial element. In general, the participatory research movement has made great strides in promoting research programmes where the ultimate users of innovations are involved and participate in the development of innovations. There are numerous cases that measure the impacts of participatory research. Whether the participatory research movement will succeed in transforming agricultural research systems so that the approach is institutionalised and its results are out-scaled to larger numbers of beneficiaries has yet to be determined. Some of the constraints to out-scaling participatory research approaches in agricultural R&D that are raised in this paper are similar to the problems of scaling up community-driven development. These difficulties include high costs, hostile institutional setting, varying levels of co-operation among different stakeholders, and lack of scaling-up logistics such as the training of large numbers of participants.

Research administrators should consider these strategies in order to promote an environment that is favourable to participatory research and will foster impact. They should also seriously consider systematic participation of farmers in research-priority setting and revise established programme mandates as necessary. Donors should consider the effects of short-term funding on the implementation of participatory research projects, which require substantial amounts of time to be spent on building and maintaining relationships and partnerships, both between farmers and researchers and among other stakeholders in the process. These considerations can accelerate the use of participatory approaches in agricultural research and development and increase the poverty-reduction dividends of agricultural research.

It seems fitting to end this set of workshop papers with an analysis of the costs and benefits of integrating participatory research and conventional research methods. For many years, participatory and conventional approaches to agricultural research have been regarded as more or less antagonistic. There is increasing evidence, however, that participatory methods can successfully be combined with conventional agricultural research. **Andreas Neef** presents experience from three sub-projects of a long-term Thai–Vietnamese–German collaborative research programme on 'Sustainable Land Use and Rural Development in Mountainous Regions of Southeast Asia', in which participatory elements were integrated into conventional agricultural research as add-on activities. In all three sub-projects, the costs of studying local knowledge or enhancing farmers' experimentation consisted of additional local personnel (facilitators, local interpreters), opportunity costs of time for participating farmers, and travel costs. However, these participatory elements of the research projects constituted only a small fraction of the total costs (2.8 per cent, 4.2 per cent, and 5.6 per cent respectively).

Hence, conventional agricultural research can be complemented by participatory components in a cost-effective way, while producing meaningful benefits in terms of (1) blending scientific and local knowledge, (2) scaling up micro-level data, and (3) highlighting farmers' constraints in adopting soil-conservation technologies.

Expanding the portfolio: practical notes on approaches and tools

A suitable point of departure is the review of **Nina Lilja and Mauricio Bellon** of the practice of participatory research at the International Maize and Wheat Improvement Center (CIMMYT). The authors take a broad look at how participatory research was implemented in 19 projects and attempt to capture some of the main lessons learned by scientists involved in these projects. The review shows that the scientists shared common lessons from their past experiences with participatory research: successful research requires dialogue and co-operation between farmers and scientists, and it is especially important that the 'problem' be defined in a collaborative and user-driven manner. However, the participatory research at CIMMYT, although a financially significant effort (US$ 9 million a year), appears to be mostly individual effort, with very little cross-project learning and interaction. Significant changes in organisational management practices and systems are often needed in order to link research more effectively with pro-poor action, or for more use to be made of the knowledge generated. It behoves CIMMYT to consider investing resources in a learning platform for scientists to share their experiences and, in doing so, add value to this research endeavour. Another critical requirement is more investment in documenting the outcomes and impacts of participatory research at CIMMYT, which would improve future applications.

Li Xiaoyun and Joe Remenyi report on the replicable framework of field-tested indicators underlying Village Poverty Reduction Planning (VPRP). Taking into account cross-cultural and institutional differences between poor communities, they develop (through a participatory research process) a participatory poverty index for use by policy makers. The paper provides insights into how participatory strategies in poverty-reduction planning can address China's chronic poverty. The authors caution that the conventional top–down measures of poverty can result in inappropriate poverty planning and interventions by local and national government, and they highlight the opportunity to improve poverty targeting significantly through the adoption of more relevant approaches, such as VPRP.

Delia Grace, Tom Randolph, Janice Olawoye, Morenike Dipelou, and Erastus Kang'ethe address the neglected area of risk management and food safety. They posit that current good practice for risk-based approaches for managing food safety require adaptation to the difficult context of informal markets; and they suggest participatory research and gender analysis as boundary-spanning mechanisms, bringing communities and food-safety implementers together to analyse food-safety problems and develop workable solutions. They give examples of how these methodologies have contributed to operationalising risk-based approaches in urban settings and are contributing to the development of a new approach to assessing and managing food safety in poor countries, which they call 'Participatory Risk Analysis'.

Ahmad Salahuddin, Paul Van Mele, and Noel P. Magor present the Poverty Elimination Through Rice Research Assistance (PETRRA) agricultural research-management project in Bangladesh which used a values-based approach in project design, planning, and implementation. Through an experiential learning process, agricultural research and development institutes, NGOs, private agencies, and community-based organisations rediscovered and improved the understanding of their strengths in meeting developmental commitment. The project successfully showed how values-based research can be meaningfully implemented and sustainable pro-poor impact achieved.

Also in terms of participatory extension, **Jon Hellin and John Dixon** report on the training of farmer-to-farmer extension agents in Peru, known locally as *Kamayoq*. In a situation where the public agricultural services were unable to serve the smallholder farmers, an NGO (Practical Action) was successful in creating a market for technical advisory services and then supplying the competent advisers (*Kamayoq*) through participatory training. This case provides some useful lessons and generalisations for future participatory research and development efforts. One of the most significant lessons is the importance of broadening the research and development framework: Practical Action began by training the *Kamayoq* in irrigation techniques, but it soon became evident that successful research/development efforts need to develop strategies that focus on strengthening linkages and effective patterns of interaction between organisations and individuals operating in the locality where impact is sought.

Jemimah Njuki, Mariam Mapila, Susan Kaaria, and Tennyson Magombo describe a process for the development of community indicators with six communities of farmers in Malawi, as part of the Enabling Rural Innovation Initiative (ERI), which is a research-for-development framework implemented by International Center for Tropical Agriculture (CIAT) and its partners. The participatory monitoring and evaluation system that they use aggregates these indicators and then uses a simple Likert scale and scoring (using a comparison of means) to apply community indicators for evaluating a community-development programme across the six communities. The paper discusses the community perceptions of the extent of the achievement of the indicators, and analyses the gender differences in perceptions. It provides an excellent practical account of community indicators of development, and an attempt is made to aggregate so that cross-community comparisons can be made.

Chanda Gurung Goodrich, Scott Justice, Stephen Biggs, and Ganesh Sah studied a rural mechanisation process in Nepal which began in the early 1990s with promotion of the use of two-wheel tractors (2WT) in order to promote minimum-tillage operations. Initial low adoption of the 2WTs was determined to be due to non-participation of the farmers in the technology-development process. The case relates the ultimate success of the project to the crucial issues of overcoming personal and institutional issues, and building coalitions among various partners and agencies through participatory technology development. This case study highlights the importance of a learning orientation in participatory research: successful research is likely to occur in systems designed for learning, rather than systems for knowing. Such programmes must often be experimental in nature – expecting and embracing failure in order to learn from it as quickly as possible.

We conclude this set of practical notes with a case study from Mozambique, by **Elisabeth Gotschi, Jemimah Njuki, and Robert Delve**, which found that gender equity is not necessarily achieved by guaranteeing men and women equal rights through established by-laws, or dealing with groups as a collective entity. An important element of this case study is the empirical evidence that the benefits of collective action in agricultural activities – for example, within farmer groups – do not necessarily promote greater gender equality in terms of improvement in women's social capital, unless multiple roles of women and men in the household and communities are taken into consideration in designing the development interventions, even if women are well represented and participate in these groups. Increasing women's leadership in these groups is a key step, but it does not address underlying power issues.

Summary of salient characteristics of the papers

A workshop on 'Rethinking Impact: Understanding the Complexity of Poverty and Change' was convened by the International Livestock Research Institute's Innovation Works programme, the PRGA Program, and the Institutional Learning and Change Initiative in March 2008. Drawing on a range of sources and from the contribution of several workshop participants, the organisers prepared a series of 21 propositions on successfully linking research with action, impact assessment, and the institutionalisation issues of the two (Kristjanson *et al.* 2008: 11–14).

Table 1: Comparison of salient characteristics of the papers and practical notes

	FE	BA	MF	TA	AA	NA	LN	LX	GD	SA	HJ	NJ	GC	GE
Linking research with action														
1. Problem definition						Y	Y		Y	Y		Y		
2. Research management										Y	Y	Y	Y	
3. Programme organisation	Y								Y	Y	Y			
4. Systems approach	Y								Y	Y				
5. Learning orientation	Y		Y	Y	Y	Y			Y	Y		Y		
6. Continuity and flexibility				Y					Y	Y	Y	Y		
7. Asymmetries of power	Y		Y								Y		Y	Y
8. Characteristics of people		Y				Y								Y
9. Broad framework	Y			Y					Y	Y	Y			Y
Evaluation, impact assessment														
1. Purpose and focus of the assessment		Y		Y		Y	Y	Y				Y		
2. Involvement of intended users	Y	Y	Y	Y		Y		Y				Y		Y
3. Impacts assessed – different approaches	Y	Y	Y		Y	Y		Y				Y		
4. Methods used and accessibility	Y	Y	Y			Y								
5. Use of assessment results					Y									Y
6. Power issues		Y	Y		Y							Y	Y	
7. Attribution issues			Y									Y	Y	
Institutions and behaviours														
1. Individual behaviour				Y		Y						Y		
2. Organisational management		Y				Y				Y		Y		
3. Policy practice				Y						Y			Y	Y
4. Knowledge-sharing practice	Y			Y		Y				Y	Y			

Notes: FE – Friis-Hansen E; BA – Bartlett A; MF – Mancini F, Jiggins J; TA – Tipilda A, Alene A, Manyong V; AA – Aw-Hassan A; NA – Neef A; LN – Lilja N, Bellon M; LX – Li X, Remenyi J; GD – Grace D, Randolph T, Olawoye J, Dipelou M, Kang'ethe E; SA – Salahuddin A, Van Mele P, Magor NP; HJ – Hellin J, Dixon J; NJ – Njuki J, Mapila M, Kaaria S, Magombo T; GC – Goodrich C, Justice S, Biggs S, Sah G; GE – Gotchi E, Njuki J, Delve R.

We have used these propositions as a guideline to summarise the 'salient characteristics of the papers and practical notes' (Table 1). This summary highlights the broad portfolio of practices presented in this special issue, and the set of papers and practical notes offer guidance on many of the key propositions (Table 1). The set is particularly strong on learning; breadth of frameworks; and design of impact assessment, especially purpose, focus, user involvement, and different approaches. In these areas, the papers document a wide range of practical experiences from which some general principles can be deduced. It should also be noted that this set of papers does not provide in-depth coverage of systems approaches or of the individual behaviour or characteristics of people. Similarly, attribution issues are not dealt with by many of the papers, perhaps reflecting the extent of existing literature on this subject.

Insights in relation to complexity science can be found in the papers dealing with systems approaches, learning orientation, continuity and flexibility, different impact-assessment approaches, and organisational management. These highlight multiple impact pathways, interaction between researchers and farmers, and feedback loops which govern the evolution of the impact pathways. From this perspective, pathway design cannot follow a static 'blueprint' of 'best practice' approach, but necessarily evolves in each situation.

Concluding remarks

This special issue presents practices and debates in participatory research and participatory technology development, concentrating on issues of implementation, impact assessment, and institutionalisation. Together these papers provide several empirical lessons for the wider development community.

Several of the papers take a fresh look at empowerment and gender. These papers illustrate that at the heart of the participatory research a fundamental question still remains: 'what are the implications of our work for the redistribution or consolidation of power among and between the world's women and men, poor and wealthy, ethnic minorities and majorities?' To better address this question, the development community, in applying participatory approaches, needs to increase its networks and reach out towards more grassroots communities, and look more broadly at economic, political, and cultural contexts and realities in order to understand what affects change.

The resources needed for participatory work are often underestimated, and methodologies for participatory monitoring and evaluation have been abundant but often not well applied. Assessing the effectiveness of participatory methods can be difficult, because they are context-sensitive; several papers in this issue review and compare the effectiveness of the impact-assessment methods, describe and analyse the use of community-based indicators, and calculate the benefits and costs of participatory research. Some of the papers focus on novel approaches to participatory research, such as construction of poverty indicators and risk assessment.

One of the most common, and important, questions about participatory research is the extent to which some of the findings generated by location-specific participatory research are applicable and transferable to similar systems elsewhere. This question is addressed in some of the papers by providing specific propositions and recommendations for the scaling-up and scaling-out efforts. That the relationship between participatory research projects and national research institutions or grassroots organisations is not always straightforward, and does not always lead to the institutionalisation of participatory approaches, is another topic discussed in many of these papers. Significant increases in scaling out participatory research approaches are unlikely unless farmers' organisations and intermediary organisations have a strong participation and presence in the research processes. Some of the lessons drawn out here are that the integration of participatory methods into differing institutional contexts requires management innovation and commitment, skill development, and new working procedures.

We hope that the broad portfolio of approaches, practices, and frameworks presented in the following 16 articles offers an operational context for better-quality implementation, evaluation, and institutionalisation of participatory approaches for the development and research community.

Acknowledgements

We would like to thank Deborah Eade for her valuable input in compiling this special issue of *Development in Practice*, and Guy Manners for his expert technical editing assistance. We also thank the organisers and participants at the 'Rethinking Impact' workshop in Cali, March 2008, for their contributions to formulating the propositions used in Table 1.

References

Ashby, J. and L. Sperling (1995) 'Institutionalizing participatory, client-driven research and technology development in agriculture', *Development and Change* 26 (4): 753–70.

Band, F. S. and K. Jax (2007) 'Focusing the meanings of resilience: resilience as a descriptive concept and a boundary object', *Ecology and Society* 12 (1): 23.

Ceccarelli, S. and S. Grando (2007) 'Decentralized-participatory plant breeding: an example of demand driven research', *Euphytica* 155 (3): 349–60.

Chambers, R. (1994) 'Participatory rural appraisal (PRA): analysis of challenges, potentials and paradigms,' *World Development* 10: 1437–54.

Chambers, R. (1995) 'Poverty and livelihoods: whose reality counts?', *Environment and Urbanization* 7 (1): 173–204.

Chambers, R. and G. R. Conway (1992) 'Sustainable Rural Livelihoods: Practical Concepts for the 21st Century', Discussion Paper 296, Brighton, UK: Institute of Development Studies.

Chambers, Robert, Arnold Pacey, and Lori Ann Thrupp (eds.) (1989) *Farmer First: Farmer Innovation and Agricultural Research*, London: Intermediate Technology Publications.

Clark, W., *et al.* (2002) *Science and Technology for Sustainable Development: Consensus Report of the Mexico City Synthesis Workshop, 20–23 May 2002*, Cambridge, MA: Initiative on Science and Technology for Sustainability. (Available at: http://www.ksg.harvard.edu/sustsci/ists/synthesis02/output/ists_mexico_consensus.pdf.)

de Jager, A., D. Onduru, and C. Walaga (2004) 'Facilitated learning in soil fertility management: Assessing potentials of low-external-input technologies in East African farming systems', *Agricultural Systems* 79 (2): 205–23.

Dickson, Nancy (2008) 'Knowledge-action Systems: The Effective Use of Knowledge to Support Decision Making', keynote paper presented at the 2008 Workshop on Rethinking Impact: Understanding the Complexity of Poverty and Change, CIAT headquarters, Cali, Colombia, 26–28 March. (Available via www.prgaprogram.org/riw/files/papers.zip.)

Dixon, J. (2004) 'Transnational learning platforms for agricultural and rural development', *Currents* 33: 30–35.

Guijt, Irene (2008) 'Seeking Surprise: Rethinking Monitoring for Collectives Learning in Rural Resource Management', PhD Thesis, Wageningen, The Netherlands: Wageningen Agricultural University.

Hellin, J., M. R. Bellon, L Badstue, J. Dixon, and R. la Rovere (2008) 'Increasing the impacts of participatory research', *Experimental Agriculture* 44 (1): 81–95.

Kauffman, S. (1996) *At Home in the Universe, the Search for Laws of Self Organisation and Complexity*, Oxford, UK: Oxford University Press.

Kristjanson, Patti, Nina Lilja, and Jamie Watts (2008) 'Rethinking Impact: Understanding the Complexity of Poverty and Change – A Pre-Workshop Dialogue. Challenge Paper', Nairobi: International

Livestock Research Institute, Cali, Colombia: PRGA Program, Rome: Institutional Learning and Change Initiative. (Available at www.prgaprogram.org/riw/files/RIW%20Challenge%20Paper-FINAL%20 edited-rev%20Jan%2025.pdf)

Lilja, Nina, John Dixon, Guy Manners, Roberto La Rovere, Jonathan Hellin, and Hilary Sims Feldstein (eds.) (2006) *New Avenues in Impact Assessment for Participatory Research*. Summary Proceedings of the Impact Assessment Workshop, October 19–21, 2005, Mexico, DF. Cali, Colombia: PRGA Program, and Texcoco, Mexico: CIMMYT.

Rogers, Patricia (2008) 'Four Key Tasks in Impact Assessment of Complex Interventions', keynote presentation to the 2008 Workshop on Rethinking Impact: Understanding the Complexity of Poverty and Change, CIAT headquarters, Cali, Colombia, 26–28 March. (Available via www.prgaprogram.org/ riw/files/papers.zip)

Snowden D. J. and M. E. Boone (2007) 'A leader's framework for decision making', *Harvard Business Review* (November 2007): 1–10.

Witcombe, J. R., R. Petre, S. Jones, and A. Joshi (1999) 'Farmer participatory crop improvement. IV The spread and impact of a rice variety identified by participatory varietal selection', *Experimental Agriculture* 35: 471–87.

Some common questions about participatory research: a review of the literature

Nina Lilja and Mauricio Bellon

This article reviews, through reference to the published literature, some key questions about participatory research. When should participatory research be used? How should participatory research be applied? What about quality of science in participatory research? Are there any institutional issues associated with the use of participatory research? And what are the benefits and costs of participatory research? The article is not a comprehensive literature review on participatory research, it is not meant to set standards for participatory research, nor to define what constitutes 'good' participatory research, but rather it seeks to summarise the realities of implementing participatory research, as discussed and debated by several published authors, and to provide some useful background for this special issue.

This article[1] reviews some overarching questions concerning participatory research that should ideally be answered before embarking upon a new research project (even better, before or during the planning of the project). These basic questions are addressed by reference to the literature. An earlier version of this article was used to identify key elements to be considered in a review of the status of participatory research at the International Maize and Wheat Improvement Center (CIMMYT) (Lilja and Bellon 2006), and was used to develop the survey questions for researchers. It is not, therefore, a complete review of the rationale for participatory research or its impacts.

The objective of participatory research

Conventional research tends to package intervention methods and programmes into one-size-fits-all, off-the-shelf approaches, based on a notion of universal best practices. Participatory methods address the drawbacks inherent in that approach by actively involving end-users in the research process, incorporating their views and representation into the prioritisation, review, conduct, and dissemination of scientific research. This fosters trust in agricultural research, increases research participation, addresses issues of greatest importance to the communities, and aids the translation of research results into useful practice.

Participatory research gives rural people within a study population opportunities to determine what is being studied, and teaches them the basics of research methodology so that they can assume collaborative roles. Furthermore, many if not most rural people in developing countries operate in imperfect markets, where prices do not completely reflect the value that these people attach to the activities that they engage in or goods that they consume or produce; therefore, simple profitability analyses – which may work well under the conditions of good market development that are common in developed countries – may be a poor guide to decision making about new activities, technologies, and products intended to improve the livelihoods of people in developing countries. Participatory research can provide a more accurate assessment of what people value that is not completely captured by market prices; this information in turn, if fed back into the design and development of new technologies, should help to make such new technologies more relevant and appropriate, so that they generate more benefits for these people. In practice, participatory approaches engage people in a community in some or all aspects of the research process – determining research questions, developing technical solutions and approaches to obtain information, and deciding what the research means and how it should be used to benefit the community.

There are two types of literature, one about types of participatory research (see, for example, Biggs 1989; Biggs and Farrington 1991; Pretty 1994), and other studies which focus on describing participatory tools and how to use them (for example, Farrington 1988; Chambers *et al.* 1989; Okali *et al.* 1994; Chambers 1997; Campbell 2001). However, there are no specific standards set for participatory research to guide research managers in monitoring what constitutes 'good' participatory research, or to help them to decide when participatory approaches would be more effective than conventional research methods, and hence to decide when their impact, via the products of the agricultural research process, on the lives of the intended beneficiaries (mostly farmers) would be more effective than that of conventional research methods.

When should participatory research be used?

Studies claim that participatory approaches are crucial in programmes that require holistic approaches (rather than changing one technology at a time), and where environmental and socio-economic conditions vary widely among farmers and sites (Roling and Wagemakers 1998). However, very few published studies provide definite decision rules based on empirical evidence for research management about when participatory approaches are most beneficial to technology development, compared with traditional centralised approaches (typically based on on-station technology development and on-farm testing). Some studies show that the traditional scientist-designed and -directed agricultural research programmes are very effective at developing varieties and technologies that can be used in farming systems that are fairly homogeneous, but often less effective when the reality of the farmer is more complex and risk-prone (Byerlee and Heisey 1996; Ohemke and Crawford 1996; Maredia *et al.* 1998; Evenson and Gollin 2002; Dalton and Guei 2003).

In reality, participatory approaches are often tried and used after the failure of conventional approaches to developing and delivering improved crop types and natural-resource management (NRM) techniques to resource-poor farmers. There are various reasons for the low uptake of agricultural technologies produced by formal research systems. Nowak (1992) defines two types of barrier to adoption: the inability to adopt, and the unwillingness to adopt. *Inability to adopt* includes situations where information about the technology is lacking; the cost of obtaining the information about it is too high; the technology is too complex or too expensive; the technology has excessive labour requirements; benefits are too far in the future; farmers have limited access to resources that might support them; managerial skills are inadequate; and farmers have no control over the adoption decision. The farmers' apparent *unwillingness to adopt* can be due to the fact that conflicting or inconsistent information is provided about the new technology; the information about the technology is difficult to apply in a farmer's particular circumstances or is irrelevant to those circumstances; there is a conflict between the current production goal and the new technology; the technology is inappropriate for the individual's physical setting; or there is an increased risk of negative outcomes. Ignorance on the part of the farmer or technology promoter, and belief in traditional practices, can also result in farmers' unwillingness to adopt new technology.

There are several studies that provide some insight into the question of when it is best to use participatory research approaches. Weltzein *et al.* (2000) use a matrix of two parameters (biophysical and economic environment) to map 65 participatory plant-breeding projects. Their biophysical environment scale ranges from high to low stress, based on actual versus expected yields, coupled with an index for incidence of crop failure. Their economic environment parameter ranges from high degree of homogeneous demand to heterogeneous demand (for instance, high-input commercial crops to low-input subsistence crops). The projects in the sample were widely dispersed within the matrix. Although many plant breeders consider participatory approaches most appropriate for environments that are high-stress and where few inputs are used (subsistence agriculture), a large number of sampled projects were located in the intermediate areas where agro-climatic stress was less severe.

Johnson *et al.* (2004) surveyed 59 participatory NRM projects, among which the most common resource across all projects was soils – nearly half of the projects worked on soil-related topics. Water was the second most common resource, followed by forests and biodiversity. The priority given to different resources varied across geographical regions. Institutional innovations were the most common technology on which projects reported working, followed by agronomic practices in Africa, and agro-forestry in Asia and Latin America. Half of the projects in the study reported working on more than one resource or technology. The average project in the inventory was developing 2.4 types of technology, directed at 1.9 types of resource.

Both of these studies also found great variations in the following factors: the objective of the research; reasons for involving various stakeholders in the research process; intended users or beneficiaries; duration of the project; geographical focus of the projects; and in other scale measures of the project.

How should participatory research be applied?

Participatory research is not an alternative research method, but an approach that can be applied to any methodology – survey, experimental, qualitative. The term 'participatory research' has sometimes been abused, creating the danger that an unrealistic ideal of participatory research may be seen as unobtainable, and may in fact discourage researchers from identifying their projects as 'engaged in participatory research'.

On the other hand, participatory research is not merely involving people more intensively as subjects of conventional research. A misconception can exist that any time of dialogue or interaction with farmers counts as 'participatory research', and scientists may identify themselves as engaged in participatory research when in fact they are only involved in contractual relationships with the farmers. As Weltzein *et al.* (2000) point out, collaborating with farmers exclusively to decentralise testing and to draw on their labour and land has nothing to do with the issue of 'participation' *per se.*

Typically, a project may contain some components that are participatory and others that are not, but to the extent that a project entails interaction between scientists and defined farmers or a group of farmers and this interaction leads to changes in the research design, technology development, or technology-diffusion pathways, one can talk of a project as 'participatory'.

There is a vast amount of literature trying to define the 'type of participatory research', or to come up with the best participatory research and gender-analysis protocol for plant breeding or NRM (for example, Martin and Sherington 1997; Weltzein *et al.* 2000; Agarwal 2001; Sperling *et al.* 2001; van der Fliert and Braun 2002; Vernooy and McDougall 2003). The reason for this is an attempt to assess which approach would have the biggest and (increasingly when we are concerned about targeting a certain type of end-user) the 'intended' impact. Many studies of types of participation reveal a variety of approaches and mixing of methods. Researchers tend to apply participatory approaches either for *functional* purposes (to increase the validity, accuracy, and particularly efficiency of the research process and its outputs) or for *empowering* purposes (to increase end-users' human and social capital) (Ashby 1996; Johnson *et al.* 2004). According to Vernooy and McDougall (2003), the objectives, scale, and scope of the research questions and local people's willingness to participate all influence the appropriateness and feasibility of the participatory research approach; some social or biophysical issues may be adequately addressed with 'low' participation, and some research issues may require 'higher' participation (for example, if the participation is intended to enable the participants to solve their own problems, such as by generating new knowledge). Less frequently, the capacity of the researchers to conduct participatory research is examined.

The type of participatory approach used also depends on local communities' or research partners' willingness and ability to participate, as well as the process through which the participants are selected and involved in the research process. Representation of community interests and local knowledge in the research process is complicated and affected, for example, by struggles over resources and gender issues. Haddinott (2002) cautions that participatory processes may enable more or less powerful groups to assert preferential rights over research outputs.

In identifying what is a 'participatory research project', an oft-cited schema has been proposed by Biggs (1989). It allows for a range of objectives for a research project, all quite valid in the right contexts. It therefore encourages a characterisation of research projects or programmes, rather than research activities considered in isolation (see Table 1). The ways in which participatory and on-station activities are ordered and co-ordinated differ considerably among projects, which bears out the idea that it is projects and programmes, and not research activities in isolation, that should be evaluated for their 'degree of participation'.

Table 1: Researcher–farmer relationships

Contract	Consultative	Collaborative	Collegial
Farmers' land and services are hired or borrowed, e.g. the researcher contracts with the farmer to provide specific types of land	There is a 'doctor–patient' relationship: researchers consult farmers, diagnose their problems, and try to find solutions	Researchers and farmers are partners in the research process and continually collaborate in activities	Researchers actively encourage the informal R&D system in rural areas

Source: Biggs (1989) as presented in Okali *et al.* (1994).

Biggs' (1989) schema is well suited as a guideline for internal assessment of a project's research methodology, but it is difficult to apply in a survey of projects to determine their degree of participation. To assess externally the type of participatory research that projects practise would require a more in-depth analysis of each project's activities, and interviews with scientists and other stakeholder participants. However, Biggs' schema suggests three key survey questions that can be used to characterise the 'type' of participatory research that projects are engaging in:

1. How were the participants (stakeholders) selected?
2. At what stage of the research did stakeholders participate?
3. What types of participatory tools were used?

Studies show that functional participatory approaches are used to improve communication between research and its clients in order to improve the design of technology and its acceptability, awareness, and adoption. This is most applicable in environments that are highly variable and difficult to manage, and it is difficult for scientific research to predict what will work where and to get clear signals from the market about farmers' and consumers' preferences that are important to technology design (Courtois *et al.* 2001; Ceccarelli *et al.* 2003; Morris and Bellon 2004). Empowering participatory approaches are used to build or enhance capacities important for beneficiaries' learning; for example, the ability to analyse opportunities, set priorities for innovations (change), seek information, experiment and draw conclusions, monitor and evaluate, and learn from mistakes. There has been little research on the conditions in which an empowering approach is most appropriate.

The quality of science in participatory research

Scientific rigour and the merits of participatory approaches have been debated in the literature (Gladwin *et al.* 2002; Hayward *et al.* 2004), due to the conventional notion of 'scientific rigour' being equivalent to replicable methods and processes. Several plant-breeding studies have formally tested the effectiveness of farmer selection *versus* breeder selection, as well as their adoption potential in terms of narrow *versus* broad adoptability (Ceccarelli *et al.* 2001, 2003; Courtois *et al.* 2001; Joshi *et al.* 2001; Joshi and Witcombe 2002).

The quality of the participatory approach used is obviously influenced by the researchers' capacity to conduct participatory research, and their views on the effectiveness and appropriateness of participatory research, which in turn are shaped by their degree of training and experience in participatory approaches, the usefulness of that training and practical experience, and their perceptions of the need to build local capacity (in the case of empowering participatory approaches).

Institutional issues

Some critics say that the advocacy of participatory research has been too prescriptive and coercive, and attention should be focused on the real impact of these methods and the receptiveness of the institutional settings in which they are advocated (Hall and Nahdy 1999). Factors related to the institutional context (such as institutional culture and practice in the planning, budgeting, and implementation of research, co-operation, and learning, and rewards and incentives to innovate), in addition to the involvement of partner institutions, will certainly influence what kinds of participatory approach it is feasible to implement (Groverman and Gurung 2001).

To assess the integration, in terms of linkages and disciplinary inclusion, of participatory projects with other projects in a particular institutional setting, it might be useful to think about linkages among scientists in terms of 'social networks'. Empirical studies of social networks show that tighter networks are actually less useful to their members than networks with loose connections to other individuals outside the main network. More open networks are more likely to introduce members to new ideas and opportunities. In other words, a group of scientists who only do things with each other already shares the same knowledge and opportunities. A group of individuals where each has connections to other social worlds is likely to have access to a wider range of information. It is better for individual project success to have connections to a variety of networks, rather than many connections within a single network.

Benefits and costs

Several studies have documented outputs of participatory research. Most of these studies use traditional indicators such as number of varietal trials, number of crosses, improvements in management techniques, number of varieties released and their potential yield-improvement effects, types of variety preferred by different types of farmer (see, for example, Heong and Escalada 1998; Snapp *et al.* 2002; Bellon *et al.* 2003; Phiri *et al.* 2004). These are important measures of the achievements of research outputs, but they are measures of success at intermediate stages. They do not quantify the impact of research on farm income, consumer welfare, or agricultural growth. Some empirical studies have captured the impact of participatory research on farm productivity and consumer welfare, and show technology adoption and rates of return calculations (Franzel *et al.* 2003; Johnson *et al.* 2003; Smale *et al.* 2003). These provide a measure of profitability of investment made in participatory activities in a given project, as compared with conventional research approaches.

The increasing use of participatory development approaches poses new challenges for decision makers and evaluators. Because these approaches are designed to be responsive to changing community needs, one of the most pressing challenges is to develop participatory and systems-based evaluation processes to allow for on-going learning, correction, and adjustment by all parties concerned. According to Hall *et al.* (2003), the greatest challenge, however, is that such holistic learning frameworks (often less quantitative in nature) must contend for legitimacy if they are to complement the dominant paradigm of economic assessment, which focuses on quantitative assessment of the rate of return on resources invested in research.

For a research-and-development institution, the value of participatory research is related to its ability to have greater impact on the lives of its intended beneficiaries through its products (typically germplasm, management practices, and policies). So it is important to place participatory research in a model to deliver impacts. It is also important to realise that, between research outputs and impacts, there is an important intermediate stage called 'project outcomes'. The 'outcomes' are the changes resulting from uses of outputs by stakeholders and clients (for example, changes in knowledge, attitudes, policies, research capacities, and agricultural practices), whereas 'impacts' are the longer-range social, environmental, and economic benefits that are consistent with the institution's mission and objectives (for example, increased agricultural productivity, improved food distribution). Ekboir (2003) argues that what counts most in research evaluation is to evaluate the new rules and patterns of participation in research networks. Ideally (and as suggested in the previous paragraph), participatory research should result in mutual learning: feedback from the end-users – not only on the products, but also on the methods – should influence the whole technology-development process. These complex interactions are frequently ignored in conventional impact-assessment studies, which focus on assessing the impact of the technology itself on end-users' livelihoods; however, knowledge of these interactions is important to those seeking to understand the full impact of participatory research.

The conventional technology-development model can be described as a linear, one-direction progression from research to outputs to outcomes and finally to impacts. A participatory model incorporates the important component of a feedback loop between the research process and outputs, and the intended beneficiaries, so that the process is adjusted to produce more relevant and appropriate outputs. The research outputs produced with participation could generate outcomes and impacts that are either similar to or different from those generated by the conventional model. Clearly, if the impacts were the same there would be little point in engaging in participation at all, unless the methodology cost less than the conventional methodology. If the outcomes are different, they could lead to 'better' or 'worse' impacts (Berardi 2002). Even if better impacts are produced, one then has to ask whether there were additional costs or savings associated with participation compared with the conventional model, and whether the benefits were worth these extra costs or savings.

Concluding remarks

Some researchers argue that the current private-sector agricultural research is narrowly focused geographically, on few crops, and on production and not consumption traits, creating a production system in which small-scale food producers will miss out on future benefits from agricultural research (Pingali 2007). This provides a justification of the importance of targeted participatory research for public-sector agricultural research. In order to target their research better to serve the needs of the poor, many public agricultural research institutions use participatory research models as a way of generating and sustaining a rapid rate of innovation, adoption, and adaptation (especially in highly uncertain and variable environments), often focusing on the crops and traits neglected by private-sector research.

Many studies and guidelines have been published on participatory research. A small fraction of that literature is represented here. This article is not meant to set standards for participatory research, nor is it meant to define what constitutes 'good' participatory research; but rather it seeks to summarise the realities of implementing participatory research, as discussed and debated by several published authors, and to provide some useful background for this special issue of *Development in Practice* by answering some 'common questions' about participatory research.

Acknowledgements

Our sincere thanks to CIMMYT for allowing us to use the material from Lilja and Bellon (2006), and to USAID for generously supporting the study through linkage funding.

Note

1. This article is based on part of Lilja and Bellon (2006).

References

Agarwal, B. (2001) 'Participatory exclusions, community forestry and gender: an analysis for South Asia and a conceptual framework', *World Development* 29 (10): 1623–48.
Ashby, J. A. (1996) 'What do we mean by participatory research in agriculture?', in *New Frontiers in Participatory Research and Gender Analysis*, Proceedings of the International Seminar on Participatory Research and Gender Analysis (PRGA), 9–14 September 1996, Cali, Colombia. Centro Internacional de Agricultura Tropical (CIAT) Publication No. 294.

Bellon, M. R., J. Berthaud, M. Smale, J. A. Aguirre, S. Taba, F. Aragon, J. Diaz, and H. Castro (2003) 'Participatory landrace selection for on-farm conservation: an example from the Central Valleys of Oaxaca, Mexico', *Genetic Resources and Crop Evolution* 50: 401–16.

Berardi, G. (2002) 'Commentary on the challenge to change: participatory research and professional realities', *Society and Natural Resources* 15 (9): 847–52.

Biggs, S. D. (1989) 'Resource-poor farmer participation in research: a synthesis of experiences from nine national agricultural research systems', *OFCOR Comparative Study Paper* No. 3, The Hague, The Netherlands: International Service for National Agricultural Research (ISNAR).

Biggs, S, and J. Farrington (1991) *Agricultural Research and the Rural Poor: A Review of Social Science Analysis*, Ottawa: IDRC.

Byerlee, D. and P. Heisey (1996) 'Past and potential impacts of maize research in sub-Saharan Africa: a critical assessment', *Food Policy* 21: 255–77.

Campbell, J. R. (2001) 'Participatory Rural Appraisal as qualitative research: distinguishing methodology issues from participatory claims', *Human Organization* 60 (4): 380–89.

Ceccarelli, S., S. Grando, E. Bailey, A. Amri, M. El-Felah, F. Nassif, S. Rezqui, and A. Yahyaoui (2001) 'Farmer participation in barley breeding in Syria, Morocco and Tunisia', *Euphytica* 122: 521–36.

Ceccarelli, S., S. Grando, R. Tutwiler, J. Baha, A.M. Martini, H. Salahieh, A. Goodchild, and M. Michael (2003) 'A methodological study on participatory barley breeding. II. Response to selection', *Euphytica* 133: 185–200.

Chambers, R. (1997) *Whose Reality Counts? Putting The First Last*, London: IT Publications.

Chambers, R., A. Pacey, and L. A. Thrupp (eds.) (1989) *Farmers First: Farmer Innovation and Agricultural Research*, London: IT Publications.

Courtois, B., B. Bartholome, D. Chaudhary, G. McLaren, C. H. Misra, N. P. Mandal, S. Pandey, T. Paris, C. Piggin, K. Prasad, A. T. Roy, R. K. Sahu, V. N. Sahu, S. Sarkarung, S. K. Sharma, A. Singh, H. N. Singh, O. N. Singh, N. K. Singh, R. K. Singh, S. Singh, P. K. Sinha, B. V. S. Sisodia, and R. Takhur (2001) 'Comparing farmers' and breeders' rankings in varietal selection for low-input environments: a case study of rainfed rice in eastern India', *Euphytica* 122: 537–50.

Dalton, T. and R. Guei (2003) 'Productivity gains from rice genetic enhancements in West Africa: countries and ecologies', *World Development* 31 (2): 359–74.

Ekboir, J. (2003) 'Why impact analysis should not be used for research evaluation and what the alternatives are', *Agricultural Systems* 78 (2): 166–84.

Evenson, R. E. and D. Gollin (eds.) (2002) *Crop Variety Improvement and Its Effect on Productivity: The Impact of International Research*, Wallingford: CAB International.

Farrington, J. (1988) 'Farmer participatory research: editorial introduction', *Experimental Agriculture* 24 (3): 269–79.

Franzel, S., C. Wambugu, and P. Tuwei (2003) 'The adoption and dissemination of fodder shrubs in Central Kenya', *ODI/AGREN Network Paper* No. 131.

Gladwin, C. H., J. S. Peterson, and A. C. Mwale (2002) 'The quality of science in participatory research: a case study from eastern Zambia', *World Development* 30 (4): 523–43.

Groverman, V. and J. D. Gurung (2001) *Gender and Organizational Change: A Training Manual*, Kathmandu: International Center for Integrated Mountain Development.

Haddinott, J. (2002) 'Participation and poverty reduction: an analytical framework and overview of the issues', *Journal of African Economies* 11 (1):146–68.

Hall, A. and S. Nahdy (1999) 'New methods and old institutions: the "systems context" of farmer participatory research in national agricultural research systems. The case of Uganda', *ODI/AgREN Network Paper* No. 93.

Hall, A., V. R. Sulaiman, N. Clark, and B. Yoganand (2003) 'From measuring impact to learning institutional lessons: an innovation systems perspective on improving the management of international agricultural research', *Agricultural Systems* 78: 213–41.

Hayward, C., L. Simpson, and L. Wood (2004) 'Still left out in the cold: problematising participatory research and development', *Sociologia Ruralis* 44 (1): 95–108.

Heong, K. L. and M. M. Escalada (1998) 'Changing rice farmers' pest management practices through participation in a small-scale experiment', *International Journal of Pest Management* 44 (4): 191–7.

Johnson, N. L., N. Lilja, and J. A. Ashby (2003) 'Measuring the impact of user participation in agricultural and natural resource management research', *Agricultural Systems* 78 (2): 287–306.

Johnson, N., N. Lilja, J. A. Ashby, and J. A. Garcia (2004) 'The practice of participatory research and gender analysis in natural resource management', *Natural Resources Forum* 28: 189–200.

Joshi, K. D. and J. R. Witcombe (2002) 'Participatory varietal selection in rice in Nepal in favourable agricultural environments – a comparison of methods assessed by variable adoption', *Euphytica* 127: 445–58.

Joshi, K. D., B. R. Sthapit, and J. R. Witcombe (2001) 'How narrowly adapted are the products of decentralized breeding? The spread of rice varieties from a participatory breeding programme in Nepal', *Euphytica* 122 (3): 589–97.

Lilja, Nina and Mauricio Bellon (2006) 'Analysis of Participatory Research Projects in the International Maize and Wheat Improvement Center', Mexico, DF: CIMMYT.

Maredia, M., D. Byerlee, and P. Pee (1998) 'Impact of food crop research in Africa', *SPAAR Occasional Papers* series, No. 1.

Martin, A. and J. Sherington (1997) 'Participatory research methods – implementation, effectiveness and institutional context', *Agricultural Systems* 55 (2): 195–216.

Morris, M. L. and M. R. Bellon (2004) 'Participatory plant breeding research: opportunities and challenges for the international crop improvement system', *Euphytica* 136: 21–35.

Nowak, P. (1992) 'Why farmers adopt production technology', *Journal of Soil and Water Conservation* 47 (1): 14–16.

Ohemke, J. F. and E. F. Crawford (1996) 'The impact of agricultural technology in sub-Saharan agriculture', *Journal of African Economics* 5 (2): 271–92.

Okali, C., J. Sumberg, and J. Farrington (1994) *Farmer Participatory Research: Rhetoric and Reality*, London: IT Publications.

Phiri, D., S. Franzel, P. Mafongoya, I. Jere, R. Katanga, and S. Phiri (2004) 'Who is using the new technology? The association of wealth status and gender with the planting of improved tree fallows in Eastern Province, Zambia', *Agricultural Systems* 79: 131–44.

Pingali, P. (2007) 'Will the Gene Revolution Reach the Poor? – Lessons from the Green Revolution', Mansholt Lecture, Waganingen University, 26 January.

Pretty, J. N. (1994) 'Alternative systems of inquiry for a sustainable agriculture', *IDS Bulletin* 25: 37–48.

Roling, N. and A. Wagemakers (1998) *Facilitating Sustainable Agriculture: Participatory Learning and Adaptive Management in Times of Environmental Uncertainty*, New York, NY: Cambridge University Press.

Smale, M., M. R. Bellon, I. M. Rosas, J. Mendoza, A. M. Solano, R. Martinez, A. Ramirez, and J. Berthaud (2003) 'The economic costs and benefits of a participatory project to conserve maize landraces on farms in Oaxaco, Mexico', *Agricultural Economics* 29: 265–75.

Snapp, S. S., H. A. Freeman, F. Simtowe, and D. D. Rohrbach (2002) 'Sustainable soil management options for Malawi: can smallholder farmers grow more legumes?', *Agriculture, Ecosystems and Environment* 91 (1/3): 159–74.

Sperling, L., J. A. Ashby, M. E. Smith, E. Weltzein, and S. McGuire (2001) 'A framework for analyzing participatory plant breeding approaches and results', *Euphytica* 122: 439–50.

van der Fliert, E. and A. R. Braun (2002) 'Conceptualizing integrative, farmer participatory research for sustainable agriculture: from opportunities to impact', *Agriculture & Human Values* 19 (1): 25–38.

Vernooy, R. and C. McDougall (2003) 'Principles for good practice in participatory research: reflecting on lessons from the field', in B. Pound, S. Snapp, C. McDougall, and A. Braun (eds.) *Managing Natural Resources for Sustainable Livelihoods: Uniting Science and Participation*, London: Earthscan Publications.

Weltzein, E., M. E. Smith, L. Meitzner, and L. Sperling (2000) 'Technical and institutional issues in participatory plant breeding from the perspective of formal plant breeding: a global analysis of issues, results and current experience', *PRGA Working Document* no. 3, Cali, Colombia: CGIAR Systemwide Program of Participatory Research and Gender Analysis for Technology Development and Institutional Innovation.

The lost 1990s? Personal reflections on a history of participatory technology development

Stephen Biggs

This article traces a history of agricultural participatory research, largely from the author's personal experience. Participatory research in the 1970s was mostly led by disciplinary scientists, and characterised by innovative activities and open academic debate, with some recognition that policy and development practice was a political process. The 1980s saw a shift to learning from past experience, and a participatory mainstream developed, seeking methods for scaling up. Meanwhile, others sought to understand and influence policy and institutional change in their political and cultural contexts, and to keep open the academic debates. The author considers the 1990s as 'lost years', during which mainstream participatory practitioners became inward-looking development generalists, not so interested in learning from others outside their paradigm. The late 2000s provide a chance to re-recognise the political and cultural embeddedness of science and technology; re-introduce strong, widely based disciplines; and learn from past activities that resulted in positive development outcomes (planned or unplanned).

Introduction: participatory technology development

In this article,[1] I use the term 'participatory technology development' (PTD) in a generic sense to include all approaches that focus on the inclusion of poor and often socially marginalised groups in helping to influence the direction and content of agricultural and natural-resources science and technology (S&T) policy practice, so that these groups are some of the main beneficiaries of S&T. It includes such approaches as farming systems research, participatory plant breeding, poverty reduction, rural mechanisation, participatory rural assessment, farmer field schools, and client-oriented innovation systems, to name but a few of the many approaches that have been and are being advocated and promoted. While I concentrate on rural technology issues, this is only as an entry point to the analysis, as the creation and promotion of 'technology' can never be dis-embedded from its historical, political, and social contexts, as some of these reflections will show.

This is a very partial story, because there are many threads to the participatory S&T narrative. I concentrate on a few of the situations in which I was involved in some way. While the lessons are derived from my own experiences, reading, and interactions with students, I hope these reflections have some resonance with the experiences of others who have worked in the arena of agriculture and natural resources, which will make the findings and conclusions more generally useful than their being anecdotal stories with little further significance.

The 1970s: interesting innovative times

The story starts with my disillusionment with quantitative macro-economic modelling as a means of investigating the long- and short-term outcomes of various Green Revolution policy options on jobs and income distribution in a highly differentiated agricultural region of India.[2] I had been one of the students at the University of California at Berkeley in the mid-1960s who had been interested in the use of Leontief input–output models and statistical methods for development planning. We thought we could predict the outcomes of different policy options for economic growth, employment, and income distribution, and provide these scenarios to 'policy makers' in governments who would make appropriate choices in a democratic way. This was part of a very technical, apolitical, and managerial way of seeing policy processes; however, my 'field experience' showed that policy and development practice was at its heart a very political and culturally defined process. Another reason for my disillusionment was that while many of the agricultural-sector models of the time claimed to look at and analyse poverty issues, when considered more carefully they did not in fact seriously model poverty and social inclusion/exclusion issues. Often only a limited set of economic policies was explored, and institutional policies relating to the distribution of assets, power and control in markets, and regulatory regimes were generally not investigated in the modelling exercises. This disillusionment with macro-policy modelling[3] led to my wanting to be more involved in policy and development practice.

From 1974 to 1976, I was seconded from the Institute of Development Studies (IDS) at the University of Sussex to work for the Ford Foundation in Bangladesh. The job involved developing projects for capacity building in agricultural and rural development research organisations, and working in on-going policy processes with the agricultural and rural development ministries. For me, some of the earliest work on participatory approaches was encountered when working with the Ministry of Rural Development, Chittagong and other universities, the Bangladesh Rural Advancement Committee (BRAC), the Association of Agricultural Voluntary Agencies, and others in Bangladesh in the early and mid-1970s. Soon after independence, many academics, bureaucrats, staff of aid agencies, and NGOs realised that there was a great number of (mostly unplanned) development initiatives taking place in the country about which little was known and from which they thought development lessons could be learned. The Ministry of Rural Development arranged for *ad hoc* teams from a range of universities and research institutes to go out, document, and understand what was happening. These teams produced 'quick reports', 'rushed reports', and 'Locally Sponsored Development Programme Series' reports, which were then reproduced and circulated by the Ministry and the research institutions involved (see, for example, Yunus and Islam 1975). Some of this research entailed interaction with the University of Sussex, a process which included critical analysis at the Science Policy Research Unit (SPRU), as illustrated by the article entitled 'The exploitation of indigenous technical knowledge or the indigenous exploitation of knowledge: whose use of what for what?' (Bell 1979) and work at IDS (Chambers and Howes 1979).

Another interesting observation from Bangladesh in the 1970s concerning the early days of participatory plant breeding (PPB) is the following quote from Brammer:

[O]ver time this has led them (farmers) to select varieties suited for their specific situations. The result is a wide diversity of rice varieties: by 1980, the Bangladesh Rice Research Institute (BRRI) had identified about 4,500 different varieties (and it considered that this might be perhaps only half of the actual number of varieties in use). This process of varietal selection is still going on. A former Director General of BRRI (Dr S. M. H. Hasanussaman) told the author that he knew of three cases where farmers had made their own selections from IR 8, one of the earliest high yielding dwarf varieties released by the International Rice Research Institute (IRRI). It is of interest to note that, in each case, those selections were made for greater plant height. That reflects the farmers' perception that the new dwarf varieties were not well adapted to situations where full water control could not be provided. To be fair, BRRI plant breeders had also quickly recognised the need for greater seedling length and plant height at maturity, and had included these plant characteristics in their breeding objectives from 1970 onwards. (2000: 116)

Another source of ideas about participatory approaches was the research of the centres of the Consultative Group on International Agricultural Research (CGIAR).[4] In the late 1970s, there was considerable CG activity in this area. I worked at the time as the regional economist for the International Maize and Wheat Improvement Center (CIMMYT) in Delhi (again on secondment from IDS). Some of the research was rationalised on the grounds that many technologies from research stations, promoted by agricultural extension agencies, were inappropriate to farmers' needs, especially the needs of poorer farmers and rural labourers in regions of agroclimatic diversity and climatic uncertainty.[5]

There was a general recognition of the need for participatory approaches to working with poorer rural households, and there was financial support from donor agencies to encourage anthropologists and economists to work with natural scientists on these issues. Many articles and documents appeared from a wide range of organisations (see, for example, Rhoades 1982; Werge 1978). I worked primarily with the all-India wheat and maize programmes on participatory breeding for triticale[6] and maize. Both of these groups were based at the G. B. Pant University at Pantnagar, Uttar Pradesh. While there were several all-India co-ordinated research programmes which had 'on-farm' trials and demonstrations, very few of these had the direct managerial involvement of senior researchers from parallel 'on-station' research activities. The Indian Secretary of Agriculture fully supported these new 'on-farm' research projects, where senior plant breeders, soil scientists, and others were substantially involved in the on-farm research.

As regards the participatory maize initiative, a number of different things happened. On the one hand, a much-respected maize breeder at Pantnagar organised an on-farm research programme of trials, surveys, and meetings, and published the findings each year, starting in 1979.[7] These reports included sections on the implications of the findings for (1) on-station maize-breeding priorities and strategies, (2) the next year's programme of on-farm research of trials, surveys, and meetings, and (3) extension recommendations. Significantly, the on-farm research reports showed how the problems of poorer households were often very different from those of richer households, and thereby questioned the prevailing research and extension strategies of recommending a general 'one size fits all' 'package of practices' and the adoption of technologies from demonstrations. These reports were important for the advocacy and promotion of participatory research methods and participatory breeding in maize, as they gave empirical evidence of how participatory methods had been effective in influencing the priorities and content of a university's maize-breeding programme, as well as acting as a basis for future on-farm research and extension advice. They were controversial at the time, as they were legitimising different ways of doing research from the mainstream national paradigm. There were attempts to have the activities stopped, and it was only because the director of the maize programme at Pantnager was a highly respected plant breeder that he was able to argue the case to continue with the on-farm research. I remember at one stage a very senior maize scientist at CIMMYT saying that the results of the on-farm research should not be published, as they were only extension activities, and that was the job of the extension ministry. I documented this case study after I left CIMMYT, because there was little research at the time that provided evidence of how participatory on-farm research had influenced breeding priorities, questioned existing extension recommendations, and provided more relevant extension advice for different socio-economic groups of farmers (Biggs 1983).

The reports of the on-farm maize programme at Pantnager continued to come out for several years, and I was told recently that they were used by other maize breeders in India in teaching and research programmes. However, after a few years, the programme lost momentum. I was told that one reason for this was that the established managerial bureaucracies of the Indian agricultural research system were very difficult to change, even if alternative approaches were available. Another reason may have been the implications of recognising that the technological problems of poorer farmers were often different from those of richer farmers. This meant that public-sector research policy would need to allocate specific funds to concentrate on these pro-poor issues. Yet another reason might have been that there were limited further productivity gains to be obtained from short-duration composite-maize breeding in that agro-climatic region at that time. It was a situation where most farmers grew monsoon maize as a crop of last resort, as it was the best crop for conditions of great climatic variability where erratic and widespread flooding was common.

Another parallel participatory maize research programme – involving maize breeders from several universities on the Gangetic plain (*terai*) – was directed at improving short-duration composite populations using integrated on-station and on-farm participatory research methods. This programme complemented the mainstream all-India maize-improvement programme. It involved improving composite varieties, using research-station plots and working with farmers' maize plots. My CIMMYT colleague in the Delhi office, the regional maize scientist, was active in this work. As far as I know, that work was never written up. This was partly as a result of the way the activities were abruptly stopped by the all-India maize programme and instructions from CIMMYT's headquarters in Mexico. This happened after a report on the participatory programme by the regional maize specialist came out in the CIMMYT Regional Report Series (Izuno 1979). This series had been circulating articles from CIMMYT and other CG Centres on participatory approaches to R&D. Some of the collateral damage of the stopping of the participatory maize-improvement programme was that the regional report series was stopped at the same time.

The other CIMMYT-related participatory breeding activity with which I worked was an on-farm triticale programme (Biggs 1980, 1982; Chauhan *et al.* 1980). In this case, the on-farm programme contributed to the termination of a breeding programme at the G. B. Pant University which had been going for about 15 years with technical support from CIMMYT and the all-India wheat programme. This outcome had been neither expected nor intended by the designers of or participants in the on-farm programme. The University breeding programme was developing triticale varieties for poor farmers in the Himalayan Hills. It was funded by the International Development Research Centre (IDRC), and a respected plant breeder from IDRC oversaw it. When the on-farm programme started in 1976, there had been no indication that the triticale-breeding programme was not producing useful materials; in fact, the opposite was the case, and the pipeline model of S&T had already got into the 'scaling up' stage with the establishment of demonstrations and the distribution of minikits of seeds and fertiliser. Unfortunately, the programme had been breeding for characteristics which were not of high priority to farmers in the Himalayan Hills. While the programme said that triticale varieties were spreading, farmer surveys failed to find any farmers who were growing triticale after the initial demonstrations and minikits. Some of the reasons for this non-adoption began to emerge at an international on-farm field workshop just before harvest. Peer-group scientists observed the on-farm trials and talked to farmers in the area. What had been planned as an international workshop to observe a 'successful and effective research programme' turned out to be a salutary lesson not only about the costs of not working closely with farmers, but also about the costs of effectively isolating research activities from adequately informed and sceptical peer-group reviewers. Whether a conventional assessment (using methods of the mainstream pipeline model) of the triticale-breeding programme would – at that time – have indicated a need for radical re-planning is hard to tell. But what can be said is that the management practices of the pipeline model in this situation had not led to effective and efficient use of research resources. The introduction of a participatory on-farm research programme had been effective – but not in the way the designers of the on-farm research activities had envisaged. At the time it was suggested that I calculate the internal rates of return to the maize and the triticale 'on-farm' research programmes. This I declined to do, as the attribution and quantitative measurement problems associated with separating out the 'on-farm research' components would have necessitated making highly questionable assumptions. In addition, I thought it more worthwhile to spend my research time investigating why and how these participatory initiatives had actually played out in practice in the political, cultural, and administrative contexts in which they took place, and learn lessons from that analysis.

The 1980s: simple approaches to 'scaling up' and 'institutionalisation'

In the 1980s, I worked a little with some participatory rice-breeding programmes on the Indian *terai*. This was a project funded by the Ford Foundation. It was a time when there were high hopes that participatory frameworks and methods developed in action–research projects would be 'scaled up' with relatively minor adaptation to national agro-climatic, economic, political, and cultural conditions. While this and other apolitical approaches to development were already being seriously challenged in the development literature,[8] the participatory mainstream – with its 'idealistic' goals – appeared rather insulated from other more broadly contextualised social-science analysis.

From my own perspective, I was interested in working with a senior rice breeder in Uttar Pradesh, who was conducting experiments with farmers on deep-water rice. He was breaking some of the conventions of the time by using large off-station pits formed at the side of railways by the removal of soil for building the railway embankments. In addition, in Bihar, senior rice breeders were conducting on-farm work with farmers. They were distributing segregating materials to farmers' fields at an earlier stage than normal in the breeding cycle. This type of research was being conducted at Patna, Cuttack, Ranchi, and near Kolkata. Some support was given to these programmes by an agricultural research policy and management course given at the University of East Anglia, where I was teaching.[9] A persistent problem at the time for Indian scientists such as these was that they were challenging more formal and mainstream conventional methods of plant breeding (see, for example, Maurya *et al.* 1988). Another issue was that some of the organisations involved in this participatory rice research in India thought that the purpose of the exercise was to devise a single new 'participatory' approach, with detailed guidelines of how it should be done, which would then be widely scaled up and promoted.

In the CG system in the 1980s, there were also some strong moves by some of the more dominant centres to 'standardise' the language and methods of participatory research. However, many centres managed to resist these moves, partly because they had developed methods which were specifically relevant to their own areas of science and empirical analysis. The agroforestry centre was particularly strong in developing participatory approaches that placed emphasis on principles rather than standard prescriptions. The 1980s was also a time when it was widely acknowledged that science did not always proceed in the orderly way that some proponents of research suggested. For example, a widely circulated 'grey' report by an anthropologist and plant breeders at IRRI revealed that farmer selection from early-generation material was already taking place at the IRRI research station. A rice-variety survey around IRRI revealed that IRRI farm labourers, some of whom were also farmers, were selecting materials from research-station experimental plots, taking them back home and doing their own informal research on their own farms. The persistent and very extensive spread of an 'escape' (Pajam/Masuli[10]) on the *terai* was another salutary reminder that the rules and procedures of conventional breeding programmes and formal release procedures were not the only source of useful, popular varieties. And today Masuli continues to be one of the most widely grown varieties in Nepal.

During the 1980s, Chambers and Ghildyal (1985) reviewed the literature on participatory and farming-systems research and adapted the 'farmer-back-to-farmer model' of Rhoades and Booth (1982) into the 'farmer-first-and-last model'. As regards assessing the 'impact' of the CG system, Chambers and Ghildyal (1985) and Chambers and Jiggins (1987) give ample empirical evidence that many of the approaches and methods of participatory research came from the innovative efforts of the CG anthropologists and economists working with natural scientists that flourished in the 1970s and early 1980s. The CG's research significantly influenced the Chambers/IDS work in the 1980s, and at least one of the major workshops on 'farmer first' was held at IDS with the involvement of a CG anthropologist (Chambers *et al.* 1989). This was followed up by another workshop in the early 1990s (Scoones and Thompson 1994).

In the mid-1980s, a research study by the International Service for National Agricultural Research (ISNAR) assessed the outcomes of about 36 participatory programmes in nine countries (see, for example, Merrill-Sands *et al.* 1991). I was a research consultant to this study of past on-farm client-oriented research (OFCOR). In some senses it was a conventional impact-assessment exercise. A research protocol was drawn up and the programmes were reviewed: what had been done well, what had gone wrong, what were successes and failures, what lessons could be learned, and what recommendations should be made to policy makers. Significantly, one of the issues that was not investigated in any depth by the study – using ethnographic/political economy tools – was the nature of the decision-making processes at the science policy, donor, government, and international levels that had led to the financing of policies and development interventions of this type.[11]

In the ISNAR research, I developed a typology for participation – contract, consultative, collaborative, and collegiate modes (Biggs 1989). This was based on an earlier framework by Ashby (1986). I used the typology to characterise overall programmes, policies, and initiatives, rather than to classify different types of trial and participation at the plant-breeding operational level, as had been Ashby's main concern. At the time, I remember that some of the more populist advocates of participatory approaches took my collegiate mode and promoted it as a goal or as a natural progression leading on from the other types. This had not been the reason for creating this typology – in fact the article had argued that different modes were relevant to different situations. I did not investigate in depth why it was that poorer people actively participated in these different modes; this was investigated by White (1996) in another context, where a similar four-part classification was used to show how and why poor people participated in these different ways.

Also during the 1980s, the Overseas Development Institute (ODI) in London had published a strong set of network papers on agriculture and natural resources. Most of these concentrated on participatory approaches, and one of the ODI papers was entitled 'Farmer-participatory research: a review of concepts and recent fieldwork' (Farrington and Martin 1987). Another participatory teaching and research programme that started in the early 1980s was the International Centre for Development Oriented Research in Agriculture (ICRA) in Wageningen. It was subsequently expanded to run in French in Montpelier (ICRA 1981). I was involved a little with this programme. One of the interesting and useful parts of this course was that it taught a wide range of participatory research approaches, and students were located as teams for an extensive field-study period in the actual day-to-day decision making of an agricultural research organisation (Mettrick 1994). In this way, the students were exposed to the political and administrative nature of research-policy practice and thus could assess the relevance of new ideas and methods.

Two other alliances that were promoting participatory approaches were the people involved with the FAO-supported integrated pest-management (IPM) programme, using experiential learning methods with farmer field schools (FFS), and the Intermediate Technology Development Group (ITDG – now called Practical Action).[12] These alliances had explicitly taken PTD into the relevant S&T political arenas in which they were working. In the case of the FFS work, Kenmore (1987) describes the way in which information from the FFS programme in The Philippines was used effectively to change pesticide policy, in spite of the power and influence of the pesticide industry in government and aid-donor circles. At the time, some applied academics were also addressing the political-economy issues of pesticide promotion and use, and asking 'do we want this to work?', because the implications of policy changes for different scientific, commercial, and other interest groups were significant (Goodell 1984).

The direct engagement of ITDG in the policy arena is illustrated by the way in which information from participatory/community development approaches used in micro-hydro-power development in Nepal was used to effectively challenge and change a joint decision by the government of Nepal and the World Bank to invest in a very large dam (Arun 3) for electricity generation (Pandey 1993). While promoters of the large dam still try to get it reconsidered, the rapid growth of a diverse micro-hydro industry has been impressive, although the equity and social outcomes have been far less than hoped for by the participatory advocates (Upadhyaya 2005).

The 1990s: critiques, questioning, and parallel discourses

In the 1990s, while some were still concerned with managerial views of developing new methods and frameworks, and scaling them up (Ashby and Sperling 1995), and others were extending the typologies of participation (Pretty 1994), there was a growing concern about the new mainstream advocacy of participatory approaches. This was expressed in critiques of the new orthodoxy (Biggs 1995)[13] and by many of the chapters in *Participation: The New Tyranny?* (Cooke and Kothari 2001). A review of some major successful poverty-reduction rural-development organisations in Asia in the early 1990s showed that most of them had tight administrative structures and little 'participation' as advocated by the participatory mainstream (Jain 1994). At this time, suggestions of alternative ways forward were central to the work based at the International Institute for Environment and Development (IIED), London. The *Participatory Learning and Action* (PLA) series was an important institutional setting for critical assessment of PTD advocacy (see, for example, Richards 1995). In parallel with the *PLA Notes*, IIED also issued a series of *Gatekeeper* case studies, which often focused on the positive outcomes of participatory projects and policies which had not arisen from an orthodox participatory approach (see, for example, Hyman 1992). Paradoxically, some of the literature from mainstream participatory researchers at this time was using a reductionist approach in the promotion of their position. They were using simple tables with two columns: in one column would be all the 'good' characteristics of participatory approaches, and in the other column would be a list of the 'bad' characteristics of 'the other' (non-participatory) approaches.[14] This was a paradox, because many of the promoters and followers of mainstream participatory themes claimed to be opposed to such types of reductionism. This type of activity did, however, promote the simplistic idea of 'them' and 'us'. This was not very useful for a more comprehensive understanding of the nature of rural poverty and technical and social change. As a teacher of Master's programmes in Development Studies (at the University of East Anglia), this often created problems for me, as there were many students who wanted frameworks and materials that presented the world in such simple ways.

An important review article by Bentley (1994) pointed to another problem that was current in the 1990s. This was the parochial and unproductive rivalries between different schools of participatory approaches. This ranged from the observation that parallel groups of researchers appeared to be self-citing each other, with little reference to, or acknowledgment of, similar work going on in other organisations, to unhelpful rivalry as regards codification, standardisation, and classification systems.

Other groups promoting participatory work with a strong gender-empowerment focus in the 1990s were ITDG and Appropriate Technology (AT) in Washington. An example of ITDG's work on gender is *Do It Herself: Women and Technical Innovation* (Appleton 1995). Also at this time, ODI's various information and newsletter networks on agriculture, forestry, and water were still operating and often addressed issues of participation, gender, and governance. The mid-1990s also saw the start of a dedicated Systemwide programme of participatory research and gender analysis (PRGA) in the CG system. This was primarily seen as a service programme to provide support to natural-resources scientists, especially plant breeders, to improve the focus and efficiency of their research.

Another independent participatory initiative begun in the early 1990s was the 'bottom–up, interactive science' approach of Bunders (2001). Bunders was a Dutch biotechnologist, and the approach was used by the overseas development ministry of the Dutch government as the central implementation framework of the new biotechnology policy in the early 1990s. This was a major radical policy change to direct Dutch biotechnology funding towards capacity development and biotechnology use in developing countries. There were programmes in Colombia, India, Kenya, and Zimbabwe. Some of the outcomes of this participatory biotechnology programme in India were written up by Clark *et al.* (2006). The formulaic action–research organisational model that the programme had used in all four countries 'worked' in India for a number of reasons: the model fitted well within the Indian bureaucratic structure; and the alliance of 'partners' in India was unusual in so far as (a) the chair of the co-ordinating committee was a very senior, highly respected, effective bureaucrat (and vice chancellor of the local agricultural university) – he had played key co-ordinating roles in several effective all-India agricultural R&D programmes; (b) the NGO partners were not 'ordinary' NGOs, but rather long-standing NGOs that had been effective politically for many years in protecting and improving the rights of labourers, poor farming households, and especially poor women; (c) the Indian biotechnology scientists were already actively engaged in interacting with poorer farmers, even before the 'bottom–up' project started, and some of their biotechnologies were already spreading; and (d) the project manager had a long-term commitment to the success of the project and stayed with it many years.[15]

In the late 1990s, the importance of relating participatory breeding approaches to regulatory and other macro-policy issues was being well argued in papers such as 'Regulatory frameworks: why do we need them?' (Farrington and Witcombe 1998). In parallel to these research and development actions, the 1990s also saw a growing literature on the anthropology of development actors. Many of these studies looked at agricultural and rural situations where participatory methods had been advocated by research and development practitioners (Long and Long 1992; Hobart 1994). An example is Lewis's (1998) ethnographic study of a fisheries project in Bangladesh, involving an international agricultural research institute and the Ministry of Fisheries. The study showed how narrow bureaucratic and organisational preoccupations overrode the participatory and development objectives of the project. Another study in Ethiopia showed how an alliance of the government of Ethiopia, environmental scientists, a big international NGO, and USAID created an environmental policy (which included the 'participation' of rural households in building terraces, etc.) that in subsequent policy evaluations was shown to have been based on little empirical evidence (Hoben 1995). Within biotechnology research, Richards (1994) showed how the social cultures of the institutions within which scientists worked influenced the direction of their research.

These ethnographic and other anthropological studies showed that simplistic and technical managerial approaches to participation were problematic, to say the least. These studies also showed that the problem was not merely a matter of 'clumsy' or badly conducted research and development, or a lack of new methods and approaches. Of course, this may have been the case in some situations; however, the ethnographic and political economy studies showed that it was the way in which participatory issues and science were framed by the participatory mainstream that was a problem. To some extent the mainstream participatory researchers were as orthodox in their advocacy and practice as the mainstream S&T actors whom they were criticising. In addition, their identity depended on their declared relationship to the more dominant S&T practice at the time.

The 2000s: recurring theme of political embeddedness of science and technology

In the late 1990s, my own work took me to Nepal, where I was seconded from the University of East Anglia to be a specialist in a large World Bank-funded agricultural research and extension project. I was based in the agricultural research council. One of the overriding issues that the scientists in the council had to contend with was the way the language and governance practice of participation had been introduced into the council since the 1970s through many, many earlier projects. While the language of participation was being used, the actual practice of science was determined by the behaviour of international and local actors. Interestingly, some of the most important changes in agricultural research-policy practice in the early 2000s came about, not as a result of a planned project with this intent, but as the result of the persistent efforts, going back 30 or more years, of Nepalese participatory researchers in a range of organisations, and their selective interaction with outside actors (see Joshi et al. 2006; Biggs 2007).

I had another recent encounter with participatory approaches during a review of groups and group-based organisations in Nepal (Biggs et al. 2005). The study investigated when and how groups and group-based organisations had been effective in bringing about institutional and policy change in terms of poverty reduction, social inclusion, and equity. Since the 1970s, a tremendous diversity of participative community-development approaches had been used in all sectors of the economy as the main 'implementing' strategy of government, donor, NGO, and development agencies. One of the study's findings was that where women's groups and alliances had been effective in policy arenas, it was a result of the specific innovative behaviour of actors (and groups of actors) in the specific political and cultural arenas in which they lived and worked. We did not find that effective innovation in policy and macro-institutional arenas came from the planned scaling up or scaling out of good or best practice of action research or participatory approaches. Rather it came about as people in different social arenas saw and/or created opportunities in the particular political and cultural contexts in which they were working.

As regards participatory plant breeding, we now have a situation where some of the ideas of PTD have been effectively reinstated in the political arena of S&T policy practice. For example, respected plant breeders in this science-policy arena are now discussing and arguing over whether one participatory breeding approach is better than another. These discussions were illustrated by the papers presented by Witcombe and by Sperling at an International Symposium on Participatory Plant Breeding and Participatory Plant Genetic Resource Enhancement in Pokhara in May 2000 (CIAT 2001).[16] If we add to this the work on rice at the West Africa Rice Development Association (WARDA) (see, for example, WARDA 2002) and that on barley at the International Center for Agricultural Research in the Dry Areas (ICARDA) (see, for example, Ceccarelli *et al.* 2001), we can say that some of the ideas of participatory breeding are being used in a number of plant-breeding programmes (Walker 2006). However, before policy recommendations are based on these experiences, it would be wise to conduct ethnographic/political economy studies to locate the activities and any lessons in their historical, political, cultural, and economic contexts.

The current decade appears to be a period when new types of social-science research are being undertaken. These studies look at the actual ways in which policy and institutional change comes about. While some of the current interest in governance is an extension of the concerns of past participatory research programmes, what is new is the re-instatement of the discipline of anthropology and the use of ethnographic and political science methods to analyse the behaviour of scientists, donors, bureaucrats, and other actors in macro-political policy arenas. The work of Lewis and Mosse (2006) and Rhoades (2006) exemplifies the new types of policy, impact, and assessment research being undertaken to understand and inform policy change. Two frequent features of this type of research are worth noting: first, that it investigates why and how things have happened in their social context (processes of social and behavioural change), thereby complementing some types of other work which attempt to define (and sometimes quantify) outcome/impact indicators; and second, that it continuously investigates the contested nature of the social relationships between different actors (especially those in the funding, policy, and macro arenas) in these on-going S&T exercises: who decides which playing field to play on, from whose perspective is it level or not, and who decides the rules of the game? In this tradition, I worked during the past few years with members of a participatory plant-breeding programme in Nepal to document and analyse how major policy changes came about in the Nepal agricultural research system (Joshi *et al.* 2006). Because these changes in policy are well supported by many years of plant-breeding research data from Nepal, published in the international arena by highly respected Nepali and outside scientists, the Nepali experience is credible. The empirical groundedness of this research means that it has a chance of influencing research policy in international agricultural organisations and national research systems. However, as the Nepal case study illustrates, whether this happens or not will depend not only on the usefulness and logic of the ideas, and on the political/cultural and economic context, but also on the institutional innovativeness of 'local' actors in the specific S&T policy arena (Biggs 2007).

What do these experiences tell us about policy practice?

Are there any lessons that can be drawn from these stories? Or are they just stories with no further usefulness? Here are some of my reflections. Others might draw other inferences from these experiences, and others may have totally different narratives about the same situations.

The lost years of the 1990s up to now?

Policy interest in PTD started in the 1970s, and there were many innovative activities in that period and healthy open academic debates about the issues raised by participation advocacy. In the 1980s, there were a number of formal reviews, assessments, and consolidations. Much of the way forward in the participatory mainstream was couched in development-intervention and managerial terms: what lessons have we learned from our past exercises, and how should we do development better? However, the mainstream did not engage in ethnographic or political-economy analysis of its own involvement in S&T practice. Some of the work of the 1970s and early 1980s revealed that participatory research itself was deeply embedded in broader political issues of national and international scientific institutions and aid donors, which in themselves were part of a broader international political economy. It was clear that the issues were not so much about the development of new toolkits of methods, frameworks, etc., *per se*, but about effective institutional innovations in the political/social arenas where S&T took place.[17] Paradoxically, some of the difficulties that the better-conceived participatory projects ran into were the result of the deeper political-economy causes of poverty, social exclusion, and inequality. However, the 'poor performance' of the projects gave some people in policy circles, who had little sympathy with engaging in the complexities of pro-poor and social equity issues, the rationale for withdrawing support and funding.

While the mainstream PTD advocates were concerned with management tools and frameworks that would be 'scaled out and scaled up',[18] two other groups of people were in a sense working in parallel to this apolitical mainstream group. One of these groups was the scientists and others who were actively engaged in the political arena of S&T. The people associated with the FAO-supported IPM programme and farmer field schools are an example: almost from the start of its activities, the IPM alliance has been effectively challenging not only conventional scientific wisdom, but also the interests of powerful actors in the S&T political arena in which pesticides and other commercial inputs are developed and used. The second parallel group has been the anthropologists, political scientists, and others who were undertaking academic studies that sought to understand how change actually took place in S&T situations, and to make suggestions for R&D intervention, based on this knowledge.

It appears that actors in mainstream participatory development often remained relatively impervious to information and 'lessons' coming from these two parallel groups; or, if they did find out about other practices, they added them to the mainstream's tool bag of frameworks and methods – but without contextualising the practices in their historical, political, cultural, and economic contextual settings. A possible reason for this was because they had become development generalists, or another reason could be that they had become too absorbed in the development sector – as if it was a separate 'sector' from other sectors of the economy. In the 1970s, many of the most innovative people were strongly trained in a discipline and applied those skills to the problems of poverty and social inclusion. In their own disciplines they could hold their own. The disciplines were pluralistic enough to encompass a wide variety of views. Another reason why some in the mainstream participatory group emphasised a managerial approach to participatory development – thereby attempting to depoliticise the issue – may have been for strategic reasons. Don't try to address powerful groups head on. Whatever the merit of such a strategy in some situations, the lost years of the 1990s mean that it was not an effective strategy in this case. Whether the 1990s are totally lost depends on whether there is evidence that mainstream participatory practitioners have learned hard lessons from the past.

Today's challenges

1. *A recognition of the political and cultural embeddness of S&T.* Today's challenges are far greater than the challenges facing people who were innovative in participatory research in the 1970s and early 1980s. With the globalisation of information and trade, causes of poverty, inequality, and social exclusion in one part of the world cannot be separated from what happens in other regions. The practice of pro-poor participatory technology development has always been embedded in the politics of science, and in the politics of policy practice more broadly. An explicit recognition of this contextualisation is now needed by those engaged in the analysis of participatory research.

2. *The re-introduction of the diversity of strong disciplines.* Associated with much of the innovative work of the 1970s were people very well trained in the broad-based rigour of their discipline, whether economics, anthropology, plant breeding, public administration, or other, and they turned their attention to the then current issues in development research. One of the greatest challenges today is how to re-introduce people who are strongly trained in the diversity of competing theories and practices of different disciplines. The need for people from such disciplines as political sciences, anthropology/ethnography, and law is particularly acute.

3. *Learn from the positive.* I think we can learn more from actor-oriented studies of situations where positive social inclusion, equity, and poverty-reduction changes have already taken place (Biggs 2007, 2008). This does not mean that we don't try to learn what we can from past 'successes' and 'failures' of research and development intervention (policies, projects, action research, etc.), but it would mean actively looking outside the 'development-intervention box' and learning from (and effectively responding to) this 'outside' information.

The mainstream participatory approaches and experiments will always be with us, and there will be funds for the development of new frameworks and methods, and for their advocacy and promotion. Some of this will be useful. However, there is plenty of evidence to show that there are other people working on participation and governance issues in their broader political sense who are following other paths, both academically and in the applied world. It is perhaps these innovators outside the participatory mainstream who are addressing today's issues of poverty and social inclusion with the innovative ability that is needed.

Acknowledgements

I am grateful to Guy Manners for much help in getting this article written.

Notes

1. This short reflective piece is an adaptation of an email to Tom Walker, who was leading a Science Council review of the CGIAR Systemwide Program on Participatory Research and Gender Analysis for Technology Development and Institutional Innovation (PRGA Program) and also undertaking a review of participatory varietal selection and participatory plant breeding for the World Bank (Walker 2006). One or two people saw the email and thought it might be interesting to others. I have slightly elaborated the original, but I hope that some of these personal reflections – partial as they are – may be of interest to others. I am grateful to Tom Walker for comments.

2. A problem for me in later years, when I had become more interested in understanding how and why change takes place, using qualitative as well as quantitative skills, was that I was sceptical when I saw people make claims for quantitative modelling and associated statistical procedures which went far beyond what the assumptions and data would allow.

3. For a discussion of Leontief input–output models, see Dorfman *et al.* (1958), and for how they can be used to investigate rural farm and non-farm livelihoods, employment, inter-sectoral linkages, and changes in the composition of aggregate consumption in a low-income country, see Falcon (1967). I described my lament for a Green Revolution policy model in an *IDS Bulletin* (Biggs 1974). I was asked to write a review article that looked at whether agricultural-sector models had seriously analysed poverty issues, and the article had an interesting history. It was commissioned by FAO for a book on agricultural-sector models, but at the last moment it was taken out and substituted by a managerial piece on how to go about good modelling practice and how policy makers should behave. However, my chapter – which showed that mainstream agricultural-sector models had not seriously addressed poverty issues – remained a ghost chapter in the book, as not all the references to it and its findings had been deleted from other parts of the book. I still do not know the politics of why and how this substitution happened. The review was published later (Biggs 1981) in a book published by the Agricultural Development Council, an organisation which no longer exists, that was concerned with strengthening disciplinary-based social-science capacities in Asia.

4. The CGIAR (or CG for short) is a strategic alliance of members, partners, and international agricultural centres whose mandate is to mobilise science to benefit the poor. The 15 international agricultural research centres supported by the CGIAR are independent institutions, each with its own charter, international board of trustees, director general, and staff. In recent years, an overall Science Council has been added.

5. For example, Perrin and Winkelmann (1976) and Farmer (1979). These articles brought to light one of the central issues in plant breeding, namely genotype–environment interaction, and the policy issues of how much money to allocate to different technology domains. Barker (1981) demonstrated how a very simple quantitative economic model, using estimates of yield changes from scientists as a result of investments in alternative breeding strategies, helped to influence research policy in IRRI at the time. Many other problems were emerging at the time from the 'Green Revolution', and these were discussed widely in the literature, as in an article by Falcon (1970) on a generation of problems.

6. Triticale is an artificial cross between wheat and rye.

7. See Agrawal *et al.* (1979); similar reports came out each year for about five years.

8. See, for example, Clay and Schaffer (1984). This type of investigation was in the tradition of earlier writers such as Hirschman (1967) who looked closely at what actually went on, and why, in development interventions.

9. At the time, we approached the International Service for National Agricultural Research (ISNAR) to see if we might work collaboratively with them in this area, but they did not see a role for any partnership or linkage with our programme. However, over the years we had close contact with the International Centre for Development Oriented Research in Agriculture (ICRA) in Wageningen.

10. Also known in Nepal as Mansuli and Mahasuri. It was a variety developed in Malaysia under a programme of the South East Asia Treaty Organisation (SEATO). It was never formally released in India, Bangladesh, or Nepal. Its name comes from Pakistan, Japan, and Malaysia, who co-operated in the programme. Hugh Brammer told me that it was probably introduced into Comilla, Bangladesh (at the time East Pakistan) by a Japanese specialist around the late 1970s. After it had spread, it was recognised formally in Bangladesh as a 'modern' variety. (The name 'high-yielding variety' (HYV) was used to designate a variety that had been through the formal breeding and official release procedures.)

11. These research-policy and macro-institutional issues are now being recognised as important, and a recent ethnographic assessment of a participatory rural development programme in India shows the embeddedness of science in the social and political context of projects, and the way in which 'success' and 'failure' of projects can depend on who decides, and when, on the criteria for the 'assessment' (Mosse 2005).

12. There were many other initiatives at the time, but these were two I had some contact with.

13. The paper was first given at a workshop on 'The Limits of Participation', organised by ITDG in 1995.

14. In discussions with Rosalind Eyben on this history, she noted that Bourdieu's framework of orthodoxy and heterodoxy might be a useful framework for investigating further the relationships between mainstream participatory advocacy and the orthodoxy that it claims to oppose.

15. The outcomes of the institutional experiments in the other countries were very different indeed. The outcomes in Zimbabwe were particularly different from what the planners of the project had in mind, as there was little evidence of any bottom–up participatory planning. However, in the late 1990s at the time of the mid-term review of the changes in the Dutch policy (I was a member of the review team), it was not clear what the project designers had expected would happen in each country, given the very different political, administrative, cultural, and aid contexts into which the institutional experimental model had been introduced.

16. Devendra Gauchan and I gave a paper at the workshop which pointed out that current mainstream participatory programmes in this area had grown out of participatory research processes that pre-dated the recent interests in the CG System and other places of the last 15 years or so (Biggs and Gauchan 2001) – hence arguing, among other things, that any 'impact' analysis of PTD projects would always have to take into account the nature of these earlier activities.

17. There are traditions of PTD in the Netherlands, France, the USA, India, and other countries. For example, in a recent review of PTD and PPB, Walker (2006) cites the history of such practices in potato research in the USA.

18. In another context, this type of activity has been called 'paradigm maintenance' (Broad 2006).

References

Agrawal, B. D., *et al*. (1979) 'Maize On-farm Research Project (1979 Report)', Pantnagar, Uttar Pradesh, India: G. B. Pant University of Agriculture and Technology.

Appleton, H. (ed.) (1995) *Do It Herself: Women and Technical Innovation*, London: IT Publications.

Ashby, J. (1986) 'Methodology for the participation of small farmers in the design of on-farm trials', *Agricultural Administration and Extension* 22 (1): 1–19.

Ashby, J. A. and L. Sperling (1995) 'Institutionalizing participatory, client-driven research and technology development in agriculture', *Development and Change* 26 (4): 753–70.

Barker, R. (1981) 'Establishing priorities for allocating funds to rice research', in *Rural Change: The Challenge for Agricultural Economists, Proceedings of the Seventeenth International Conference of Agricultural Economists held at Banff, Canada 3rd–12th September 1979*, Aldershot: Gower.

Bell, M. (1979) 'The exploitation of indigenous technical knowledge or the indigenous exploitation of knowledge: whose use of what for what?', *IDS Bulletin* 10 (2): 44–50.

Bentley, J. W. (1994) 'Fact, fantasies and failures of farmers' participatory research', *Agriculture and Human Values* 11 (2&3): 140–150.

Biggs, S. D. (1974) 'Lament for policy oriented research: observations on a research project to formulate a computer model for regional rural planning in the Kosi Region, Bihar, India', *IDS Bulletin* 5 (4): 28–36.

Biggs, S. D. (1980) 'On-farm research in an integrated agricultural technology development system: a case study of triticale for the Himalayan Hills', *Agricultural Administration* 7 (2): 133–45.

Biggs, S. D. (1981) 'Agricultural models and rural poverty', in Max R. Langham and Ralph H. Retzlaff (eds.) *Agricultural Sector Analysis in Asia*, Singapore: University of Singapore Press.

Biggs, S. D. (1982) 'Generating agricultural technology: triticale for the Himalayan Hills', *Food Policy* 7 (1): 69–82.

Biggs, S. D. (1983) 'Monitoring and control in agricultural research systems: maize in Northern India', *Research Policy* 12: 37–59.

Biggs, S. D. (1989) 'Resource-poor farmers participation in research: a synthesis of experiences from nine national agricultural research systems', *OFCOR Comparative Study Paper 3*, The Hague: ISNAR.

Biggs, S. D. (1995) 'Participatory technology development: a critique of the new orthodoxy', in *Two Articles Focusing on Participatory Approaches, AVOCADO Series 06/95*, Durban: Olive Organisation Development and Training.

Biggs, S. D. (2007) 'Reflections on the Social Embeddedness of S&T in Rural and Agricultural Transformations: Learning from Positive Experiences of Poverty Reduction and Social Inclusion', paper prepared for the International Conference on Policy Interventions and Rural Transformations: Comparative Issues (hosted by the College of Humanities and Development of the China Agricultural University) Beijing, 10–16 September.

Biggs, S. (2008) 'Learning from the positive to reduce rural poverty and increase social justice: institutional innovations in agricultural and natural resources research and development', *Experimental Agriculture* 44 (1): 37–60.

Biggs, S. and D. Gauchan (2001) 'The broader institutional context of participatory plant breeding in the changing agricultural and natural resources research system in Nepal', in CIAT (2001).

Biggs, Stephen D., Sumitra M. Gurung, and Don Messerschmidt (2005) 'An Exploratory Study of Gender, Social Inclusion and Empowerment through Development Groups and Group-Based Organisations in Nepal: Building on the Positive', Report submitted to the Gender and Social Exclusion Assessment (GSEA) Study, National Planning Commission, World Bank and DFID, Kathmandu (Version 2, March 2005), available at http://www.prgaprogram.org/IAWFTP/IA%20WEB/resourcess.htm (retrieved 26 October 2007).

Brammer, H. (2000) *Agroecological Aspects of Agricultural Research in Bangladesh*, Dhaka: University Press.

Broad, R. (2006) 'Research, knowledge, and the art of "paradigm maintenance": the World Bank's Development Economics Vice-Presidency (DEC)', *Review of International Political Economy* 13 (3): 387–419.

Bunders, J. (2001) 'Utilization of technological research for resource-poor farmers: the need for an interactive innovation process', in *Utilization of Research for Development Cooperation: Linking Knowledge Production to Development Policy and Practice (Publication No. 21)*, The Hague: Netherlands Development Assistance Research Council (RAWOO).

Ceccarelli, S., S. Grando, E. Bailey, A. Amri, M. El-Felah, F. Nassif, S. Rezqui, and A. Yahyaoui (2001) 'Farmer participation in barley breeding in Syria, Morocco and Tunisia', *Euphytica* 122: 521–36.

Chambers, R. and B. P. Ghildyal (1985) 'Agricultural research for resource-poor-farmers; the farmer-first-and-last model', *Agricultural Administration* 20: 1–30.

Chambers, R. and M. Howes (1979) 'Rural development: whose knowledge counts?' *IDS Bulletin* 10 (2): 5–11.

Chambers, R. and J. Jiggins (1987) 'Agricultural research for resource poor farmers. Part 1: Transfer-of-technology and farming systems research', *Agricultural Administration and Extension* 27 (1): 35–52.

Chambers, R., A. Pacey, and L. A. Thrupp (eds.) (1989) *Farmer First: Farmer Innovation and Agricultural Research*, London: IT Publications.

Chauhan, K. P. S., M. G. Joshi, *et al.* (1980) 'Operational Research Project on Triticale at Pantnagar: Final Report, September 1, 1975 to 31 December, 1979', Pantnagar, Nainital, India: G. B. Pant University of Agriculture and Technology, Directorate of Experimental Station.

CIAT (2001) *An Exchange of Experiences from South and South East Asia* (Proceedings of the International Symposium on Participatory Plant Breeding and Participatory Plant Genetic Resource Enhancement, May 1–5, 2000, Pokhara, Nepal), Cali, Colombia: Participatory Research and Gender Analysis, International Center for Tropical Agriculture (CIAT).

Clark, N., P. Reddy, and A. Hall (2006) 'Client-driven Biotechnology Research for Poor Farmers: a Case Study from India', paper prepared for the workshop on Governing Technology for Development: From Theory to Practice and Back Again, organised by the Open University, 31 March to 1 April.

Clay, E. J. and B. B. Schaffer (eds.) (1984) *Room for Manoeuvre: An Exploration of Public Policy in Agriculture and Rural Development*, London: Heinemann.

Cooke, B. and U. Kothari (eds.) (2001) *Participation: The New Tyranny?* London: Zed Books.

Dorfman, Robert, Paul A. Samuelson, and Robert M. Solow (1958) *Linear Programming and Economic Analysis*, New York, NY: McGraw Hill.

Falcon, W. P. (1967) 'Agricultural and industrial inter-relationships in West Pakistan', *American Journal of Agricultural Economics* 49 (5): 1139–54.

Falcon, W. P. (1970) 'The green revolution: generation of problems', *American Journal of Agricultural Economics* 52 (5): 698–709.

Farmer, B. H. (1979) 'The Green Revolution in South Asian rice fields: environment and production', *Journal of Development Studies* 15 (4): 304–19.

Farrington, J. and A. Martin (1987) 'Farmer-participatory research: a review of concepts and recent fieldwork', *Agricultural Administration and Extension* 29: 247–64.

Farrington, J. and J. R. Witcombe (1998) 'Regulatory frameworks: why do we need them?', in J. R. Witcombe, D. S. Virk, and J. Farrington, *Seeds of Choice: Making the Most of New Varieties for Small Farmers*, London: IT Publications.

Goodell, G. (1984) 'Challenges to international pest management research and extension in the Third World: do we really want IPM to work?' *Bulletin of the Entomological Society of America* 30 (3): 18–26.

Hirschman, A. O. (1967) *Development Projects Observed*, Washington: Brookings Institute (reprinted in 1995).

Hobart, M. (ed.) (1994) *An Anthropological Critique of Development: The Growth of Ignorance*, London: Routledge.

Hoben, A. (1995) 'Paradigms and politics: the cultural construction of environmental policy in Ethiopia', *World Development* 6: 1007–22.

Hyman, E. L. (1992) 'Local agro-processing with sustainable technology: sunflower seed oil in Tanzania', *Gatekeeper Series No. 33*, London: IIED.

ICRA (1981) 'Agricultural research and its clientele: report on the ICRA experimental course 1981', *Bulletin* no. 1, Wageningen: International Centre for Development Oriented Research in Agriculture (ICRA).

Izuno, T. (1979) 'Maize in Asian cropping systems: historical review and considerations for the future', *CIMMYT Asian Reports No. 9*, New Delhi: CIMMYT Asian Regional Office.

Jain, P. S. (1994) 'Managing for success: lessons from Asian development programs', *World Development* 22 (9): 1363–77.

Joshi, K. D., S. Biggs, D. Gauchan, K. P. Devkota, C. K. Devkota, P. K. Shrestha, and B. R. Sthapit (2006) 'The evolution and spread of socially responsible technical and institutional innovations in a rice improvement system in Nepal', *Discussion Paper 8*, Bangor, UK: CAZS-Natural Resources, University of Wales.

Kenmore, P. E. (1987) 'Crop loss assessment in a practical integrated pest control program for tropical Asian rice', in P. S. Teng (ed.) *Crop Loss Assessment and Pest Management*, St Paul, Minnesota: American Phytopathological Society Press.

Lewis, D. J. (1998) 'Partnership as process: building an institutional ethnography of an inter-agency aquaculture project in Bangladesh', in D. Mosse, J. Farrington, and A. Rew (eds.) *Development as a Process: Concepts and Methods for Working with Complexity*, London: ODI, and New York: Routledge.

Lewis, D. and D. Mosse (eds.) (2006) *Development Brokers and Translators: The Ethnography of Aid and Agencies*, Bloomfield, CT: Kumarian Press.

Long, N. and A. Long (eds.) (1992) *Battlefields of Knowledge: The Interlocking of Theory and Practice in Social Research and Development*, London: Routledge.

Maurya, D. M., A. Bottrall, and J. Farrington (1988) 'Improved livelihoods, genetic diversity and farmer participation: a strategy for rice breeding in rainfed areas of India', *Experimental Agriculture* 24: 311–20.

Merrill-Sands, D., S. D. Biggs, R. J. Bingen, P. T. Ewell, J. L. McAllister, and S. V. Poats (1991) 'Integrating on-farm research into national agricultural research systems: lessons for research policy, organization and management', in R. Tripp (ed.) *Planned Changes in Farming Systems: Progress in On-farm Research*, Chichester: John Wiley & Sons.

Mettrick, H. (1994) *Development Oriented Research in Agriculture: An ICRA Textbook*. Wageningen: International Centre for Development Oriented Research in Agriculture (ICRA) and Technical Centre for Agricultural and Rural Cooperation (CTA).

Mosse, David (2005) *Cultivating Development: An Ethnography of Aid Policy and Practice*, London: Pluto Press.

Pandey, P. (1993) 'Closing down options for hydro-power in Nepal', *Appropriate Technology* 20 (3): 4–6.

Perrin, R. K. and D. Winkelmann (1976) 'Impediments to technical progress on small versus large farms', *American Journal of Agricultural Economics* 58 (5): 888–94.

Pretty, J. N. (1994) 'Alternative systems of inquiry for sustainable agriculture', *IDS Bulletin* 25 (2): 37–48.

Rhoades, R. E. (1982) 'The art of the informal agricultural survey', *Social Science Department Training Document 1982-2*, Lima: International Potato Center.

Rhoades, R. E. (2006) 'Seeking half our brains: reflections on the social context of interdisciplinary research and development', in M. M. Cernea and A. H. Kassam (eds.) *Researching the Culture in Agri-Culture: Social Research for International Development*, Oxford: CABI Publishing.

Rhoades, R. E. and R. H. Booth (1982) 'Farmer-back-to-farmer: a model for generating acceptable agricultural technology', *Agricultural Administration* 2: 127–37.

Richards, P. (1994) 'The shaping of biotechnology: institutional culture and ideotypes', *Biotechnology and Development Monitor* (18 March): 24.

Richards, P. (1995) 'Participatory rural appraisal: a quick and dirty critique', in *PLA Notes Number 24*, London: IIED.

Scoones, I. and J. Thompson (eds.) (1994) *Beyond Farmer First: Rural People's Knowledge, Agricultural Research and Extension Practice*, London: IT Publications.

Upadhyaya, S. K. (2005) 'Benefit sharing from hydropower project in Nepal', *Equitable Hydro Working Paper 6*, Kathmandu: Winrock International.

Walker, T. S. (2006) *Participatory Varietal Selection, Participatory Plant Breeding, and Varietal Change*, Washington, DC: World Bank.

WARDA (2002) *Participatory Varietal Selection: Beyond The Flame*, Bouaké, Côte d'Ivoire: West Africa Rice Development Association (WARDA).

Werge, R. (1978) 'Social science training for regional agricultural development', *CIMMYT Asian Reports No. 5*, New Delhi: CIMMYT Asian Regional Office.

White, S. (1996) 'Depoliticising development: the uses and abuses of participation', *Development in Practice* 6 (1): 6–15.

Yunus, M. and M. Islam (1975) 'A report on Shahjalaler Shyamal Sylhet', *Locally Sponsored Development Programme Series: Report No. 1* (Rural Studies Project, Department of Economics, Chittagong University, Bangladesh), Dhaka: Ministry of Cooperatives, Local Government and Rural Development.

Impact assessment of farmer institutional development and agricultural change: Soroti district, Uganda

Esbern Friis-Hansen

This article is based on participatory development research conducted in Soroti district of Uganda with the aim of assessing the impact of agricultural development among poor farmers. The central argument is that a combination of farmer empowerment and innovation through experiential learning in farmer field school (FFS) groups, changes in the opportunity structure through transformation of local government staff, establishment of new farmer-governed local institutions, and emergence of a private service provider has been successful in reducing rural poverty. Based on an empirical study of successful adaptation and spread of pro-poor technologies, the study assesses the well-being impact of agricultural technology development in Soroti district. The study concludes that market-based spread of pro-poor agricultural technologies requires an institutional setting that combines farmer empowerment with an enabling policy environment.

Introduction

Poverty prevails in sub-Saharan Africa, where the proportion of people living in poverty remained stable at about half the region's population between 1981 and 2001. However, because of population growth, the number of poor people almost doubled from 164 million in 1981 to 316 million in 2001. Poverty is largely a rural phenomenon and is often more severe in rural areas than in urban centres. In East and Southern Africa, rural areas with high potential for production and marketing accommodate the majority of poor people, while poverty is more severe in semi-arid or otherwise marginal rural areas (IFAD 2002).

After a decade of involvement by the Consultative Group on International Agricultural Research (CGIAR) and national agricultural research systems (NARS) in participatory agricultural technology development processes, questions have rightly been asked about its effectiveness and impact on alleviating poverty (Egelyng 2005). While participatory technology development has without doubt proved a sound way of enhancing the relevance of agricultural technology for the participating farmers, its wider impact on reducing rural poverty is less clearly documented.

The study on which this article is based was undertaken in response to this gap in knowledge and available methodologies. The study sought to understand the institutional context that enables agriculture technology development and poverty reduction among smallholder farmers. On the one hand, the study examined the changes in opportunity structures – that is, the responsiveness, quality, and relevance of agricultural advisory services provided by the public and private sectors. On the other hand, it analysed changes in farmer empowerment – that is, the ability of farmers to effectively articulate informed demands for advisory services.

Apart from project-based monitoring and evaluation studies, few comprehensive impact-assessment activities have been conducted in association with the spread of pro-poor technology in East and Southern Africa. The understanding of poverty in these impact-assessment studies is furthermore often instrumental, providing an inadequate understanding of technology-adoption processes and pathways out of poverty.

This study analysed the socio-economic and institutional context in which pro-poor technologies are adopted by poor farmers, and assessed its impact on well-being among members and non-members of farmer groups.

Development of smallholder agriculture in Soroti district

Fieldwork in Soroti district, Uganda, was undertaken in 2001 and 2004. Smallholder agricultural development in the district had experienced considerable success over the previous decade, in spite of the fact that it is characterised by unstable rainfall and poor soil fertility.

Soroti district, in Eastern Uganda, has been a test bed for many agricultural development initiatives. The district has a land area of 3715 km^2; it is traversed by numerous swamps and other ravine wetlands. Annual rainfall is between 1100 and 1200 mm, but unreliable rains lead to droughts and floods. The soils are, to a large extent, poor, shallow, and light-textured with high sandy loam content. The traditional Teso farming system, which is prevalent in Soroti, Kumi, Katakwi, and Kaberamaido districts, supports a wide range of cash and food crops, placing the districts where the system is used among the highest agricultural performers in Uganda (Parsons 1970).[1] During the 1980s, agriculture was depressed by civil war; dramatic de-stocking as a result of major cattle raids by pastoralists during the late 1980s and early 1990s further undermined agricultural development. Smallholder agriculture in Soroti district has, however, experienced growth since the mid-1990s. The subsequent decade was characterised by peace, improved access to market opportunities, and improvements in local government structures, including reform of the institutional context influencing smallholder agriculture. Nevertheless, farm incomes are still low, so access to new markets and technological innovation are key elements in reducing rural poverty.

Using literature and discussions with key informants among local government (LG) and farmers, the study identified three policies and programmes that have strongly influenced the rural institutional context that underpins smallholder agriculture in Soroti district: (1) the Local Government Act 1997; (2) the Farmer Field School programme 1999–2002; and (3) the National Agricultural Advisory and Development Services (NAADS) programme initiated in 2002. The implications of these policies for changes in opportunity structures and farmer empowerment will be examined in the following.

Framework for analysing smallholder agricultural development in Soroti district

Empowerment enables people to influence decision processes and undertake transformative actions which help them to improve their livelihoods (*Voices of the Poor* trilogy: Narayan *et al.* 2000a, 2000b; Narayan and Petesch 2002). At the same time, the contemporary use of 'empowerment' seeks to identify power in the capacity of people to increase their self-reliance and individual strengths, rather than in terms of a more political concept that stresses the relations between individuals and between groups. The study adopted the following definition of farmer empowerment: '*A process that increases the capabilities of smallholder farmers and farmer groups to make choices and to influence collective decisions towards desired actions and outcomes on the basis of those choices*' (Friis-Hansen 2004b).

Knowledge and organisation have been identified as the most important aspects of farmer empowerment. Knowledge-empowerment enables farmers to understand the causes and effects of their own agricultural problems and to articulate their needs in terms of technology, extension, and development as informed demands. Knowledge-empowerment allows farmers to participate actively in the planning, implementation, and evaluation of services, in effect transforming them into clients, managers, and/or owners/partners, rather than passive beneficiaries. Organisational empowerment is realised when farmers are organised in groups that are coherent, independent, and sustainable. Such groups can enable farmers to articulate their informed demands and interact with state institutions and the private sector. They are also the basis for joining/establishing higher-level farmers' organisations that could represent their interests at local government and national policy levels.

While a farmer's individual capabilities are important factors in achieving improvements in his/her livelihood situation – for example, increased income from production, improved service provision – they do not alone determine these development outcomes. These are also dependent upon the conditions present for engaging in production, for accessing services and resources, and for controlling assets. Such conditions are structured by the policies, rules, and practices found in social and economic institutions, and not least by the policies of the government. These we term 'opportunity structures',[2] which provide the context in which farmers act and influence the development outcomes achieved (Friis-Hansen 2004b). Opportunity structures tend to be structural in nature and institutionalised in terms of law, cultural and social practices, and economic and political interests.

Changes in opportunity structures in Soroti district

The three policy changes identified in the previous section influenced the opportunity structures of smallholder agriculture in Soroti district in the following ways: (1) enhanced responsiveness of extension services; (2) institutional transformation of LG Agricultural Staff; and (3) emergence of private-sector agricultural service providers. These changes in opportunity structures are briefly discussed below.

Institutional transformation of local government extension staff Soroti district, like all districts in Uganda, was decentralised in accordance with the Local Government Act 1997.

Political and financial powers have been devolved to district and sub-county levels, bringing services nearer to rural people. Central government's roles were largely reduced to policy formulation, co-ordination, standardisation, and regulation of services.

However, unlike most other districts, Soroti extension department viewed decentralisation as a chance to gain independence from the top–down centralised Training and Visit (T&V)-inspired unified extension system which treated farmers as passive recipients of externally formulated technology packages in the form of extension messages and demonstrations (Friis-Hansen 2004a). In 1996 the department of extension started a process of institutional transformation which aimed to change attitudes and modes of operation that were rooted in the conventional T&V extension method. This independence appears to have stimulated innovation among extension staff, with (according to extension staff interviewed during the study) resultant designing of some crude but workable farmer-managed programmes which began to empower farmers to advocate for their development rights through participatory bottom–up planning processes.

With support from a Dutch-funded client-driven extension project, district-level public extension staff were assessed by farmers and rewarded accordingly by management. Parish Agricultural Development Committees (PADEC), each comprising five locally elected farmers, functioned as 'para-extensionists' and interacted with local government field extension workers (FEW). Through a mix of 'carrots' and 'sticks', the District Agricultural Officer encouraged a gradual change in the relationship between PADEC and FEW, with PADEC taking on more tasks and the FEW being increasingly held accountable to the PADEC. FEW were given logbooks indicating their tasks and activities, which should be assessable by PADEC during working hours. FEW were to earn incentives according to their performance, monitored through a point-scoring system relating to activities completed and other indicators such as the logbook, and review by PADEC and exam results of participatory training sessions (interview with extension staff).

Responsiveness of advisory services The NAADS programme in Soroti is being implemented under the Uganda government policy of decentralisation. Soroti is a decentralised district with 14 rural sub-counties and one municipal council. Each sub-county is a decentralised unit of governance able to plan and mobilise resources for its development activities. NAADS is currently being implemented in 13 sub-counties.

NAADS is based on farmer groups managed through farmer representatives at sub-county and district levels known as 'farmer fora'. The sub-county farmer fora consist of 15 members, including a chair, a secretary, and a procurement sub-committee of seven members. The district farmer fora are made up of the chairs of the sub-county farmer fora. Likewise, the national farmers' forum draws representation from the district chairpersons. NAADS is managed at the national level by a secretariat and a board, overseen by the Ministry of Agriculture, Animal Industry and Fisheries (MAAIF). At the district and sub-county levels, District and Sub-County NAADS Co-ordinators co-ordinate the programme. Sub-County and District Councils monitor, supervise, and guide the operations of the programme.

Technology generation, enterprise development, and market linkages are key outputs of NAADS. The key components include advisory and information services to farmers; development of private-sector institutional capacity; improving the programme management capacity; assuring the quality of services delivered; and improving technology and market linkages for farmers.

With the establishment of sub-county farmer fora, members of NAADS groups gained (in principle) an opportunity to articulate knowledge and technology needs effectively and to influence the selection of agricultural service providers. In practice, an NGO is hired by the District NAADS co-ordinator to facilitate the sub-county farmer fora to identify needs and select enterprises. A review of progress reports from these NGOs during 2002–2004 revealed that a tedious process of needs articulation is undertaken, in which all NAADS groups express their knowledge and technology needs, which are consolidated at Parish level and presented during a sub-county farmer fora meeting, in which the facilitator and district NAADS co-ordinator participate. However, the review of progress reports also revealed that the decisions taken at this meeting regarding enterprise selections often disregard the needs articulated by the various NAADS groups. The reason given is that the NAADS implementation manual, issued by the NAADS secretariat in Kampala, states that the selected enterprises have to be 'commercial'. This has been interpreted by the NGOs and NAADS co-ordinator to mean that only purely commercial enterprises, such as bee-keeping, cultivation of cotton, establishment of fruit trees, and such like can be supported. The result has been that many, if not most, of the knowledge needs articulated by NAADS groups have been disregarded. The most popular need expressed is, for example, technologies to improve cultivation of cassava. However, cassava is not regarded as a commercial crop.

The study thus shows that the process of enterprise selection is systematically biased towards technical advice for commercial enterprises. This limits the responsiveness of service providers to the knowledge needs articulated by NAADS groups, which causes frustration among farmers and may, if allowed to continue, undermine farmers' trust in the demand-driven advisory-service model.

Emergence of private-sector service providers under NAADS A highly supportive LG in Soroti district has been conducive to the establishment of an enabling environment for the establishment of private-sector services in response to new opportunities provided by the NAADS policy. Twenty-five private agricultural advisory companies emerged in Soroti district between 2002 and 2007, all relatively small (with fewer than 15 employees). They typically carry out short-term contracts of a few months' duration at a time for sub-county farmer fora. The NAADS District Co-ordinator assists the sub-county farmer fora in formulating terms of reference, which are displayed on the notice board of the district NAADS office and tendered to the private service providers. There seems to be healthy competition among the different service-providing companies, although they all have capacity limitations and are able to conduct only a limited set of assignments in a given period (interview with NAADS co-ordinator).

The sub-county farmer fora show a tendency to place increasing demands on the quality of work carried out by the private service providers, which is monitored by a sub-committee of the sub-county farmer fora. This indicates that the participatory monitoring system is proving efficient. The simple monitoring reports give several examples of cases where farmers' disappointment over the services provided led to changed behaviour by the service provider, or in some cases termination of contract.

Interviews with employees of the private service providers paint a picture of general satisfaction among staff. In particular, young extension graduates working for private service providers expressed satisfaction in having to respond to farmers rather than to the district extension officer.

Farmer empowerment and innovation

The farmer field school (FFS) approach was introduced into the district between 1999 and 2002 under the East African small sub-regional pilot project for farmer field schools (financed by International Fund for Agricultural Development [IFAD] and implemented by the Global Integrated Pest Management Facility Project under the auspices of FAO).

The objectives of FFS in Soroti district include the following: (1) to increase the expertise of farmers to make logical decisions on what works best for them, based on their own observations of experimental plots in their FFS; (2) to establish coherent farmer groups that facilitate the work of extension and research workers, providing the impetus for a demand-driven system; (3) to enhance the capacity of extension staff to serve as technically skilled and group-sensitive facilitators of farmers' experimental learning. Rather than prescribing blanket recommendations that cover a wide geographic area, the methods train the extensionists to work with farmers in the validation and adoption of methods and technologies.

By 2002, when the FFS project ended, some 192 FFSs had been established in Soroti district, following a foci model, with at least 15 FFSs living in relative proximity to each other in each sub-county. This model ensured collaboration and synergy among FFS members. About 4800 farmers underwent season-long training in integrated production and pest management (IPPM). Of these, 90 farmers underwent a refresher training of trainers to become farmer-facilitators to establish FFSs in their respective sub-counties.

The FFS approach exposed farmers to a learning process in which small groups (four or five farmers) regularly observe a field as an entire ecosystem and learn to make crop-management decisions based on an analysis of their observations. Farmers' capacity to validate new technologies gradually improves in FFS groups, which over time provides FFS group members with multiple ways of responding to field situations.

The systematic season-long training following the crop cycle from land preparation to harvest enables the farmers to adapt technologies to suit their situation and to become more responsive to change. The methodology has proved effective in group formation and motivation, and in enabling farmers to undertake farming-oriented self-learning with a trained moderator (Duveskog 2006).

Integrated pest and production management (IPPM) was the entry point when FFS started in Soroti in 1998. Farmers' priorities have since then added a range of other issues to the FFS curriculum, in particular issues of market access and quality standards. The most important additions are HIV and AIDS, basic principles of nutrition, reproductive and family health care, malaria control, immunisation, basic principles of environmental management, water and soil conservation, and basic financial management skills. The multi-dimensional approach has led to strong informal links among government departments, NGOs, community-based organisations (CBOs), researchers, and other service providers. This has been made easier by the grant system used in the programme. At the establishment of the FFS, farmers, under the guidance of a facilitator, write a simple grant proposal, stipulating their background, common goal, what they intend to do, their contribution, sustainability of the group, work plans, and budget for the season-long training. Then funds are transferred directly to their bank account, including money to pay the facilitators' allowances.

Qualitative interviews with FFS members and leaders indicated that participation in a season-long learning cycle has greatly improved FFS members' analytical skills and enabled them to articulate demands more accurately and effectively. (See Box 1 for an example.) A second effect, mentioned in all the interviews, was the creation of trust among FFS group members. Even though external support for the FFS groups ended in 2002, many of the groups have continued to function, using their own savings to finance activities.

A clear indicator of the strong farmer-empowerment aspects of the FFS approach to learning and organisation is its positive effect on the establishment of NAADS groups and farmer fora. There was a marked difference in the pace at which fora were established in the different sub-counties and in how well they function. In sub-counties where a critical mass of FFS groups existed, these seized the opportunity and converted into NAADS groups. Interviews with sub-county farmer fora members indicated that individual FFS graduates who no longer belonged to an FFS group had often been the driving force in establishing new NAADS groups.

Box 1: Summary of lifecycle interview with Mrs Grace Asio, Chairperson for Asureli women's FFS

Together with 29 other women, I took part in forming a new women-only FFS in 2000. None of us were well off at the time we joined the group. During the first year of FFS, we went through a classic FFS curriculum, studying the growth of cotton plants, and associated pest and diseases and IPM solutions, including identification of insects and timing of spraying. During the second year, we shifted to groundnuts and acquired new knowledge of insects of and natural pesticides for this crop. We learned to use changes in appearance of the groundnut leaves as an indicator of plant health. During the third year, we experimented with spraying pesticides made from the neem tree and compared it with use of chemical pesticides (no major difference in effect). During our year four (the current season), we are discussing market outlets for cash crops, including groundnuts, sunflower, and sweet potatoes. Our FFS group has continued to cultivate a common field, but all members have also applied what we have learned in our individual fields. During weekly group meetings, we inspect each other's fields. Our group has a bank account in which part of the proceeds from sale of crops from our common field is accumulated.

Changes in my life since joining the FFS/NAADS:

(i) I have gained more confidence and am now able to speak out my demands to service providers. (ii) I have learned better crop management and use it in my fields. (iii) I have changed my mind-set and now understand the importance of agriculture as a business. (iv) I have become better at experimenting, innovating, and generating new ideas. (v) I have learned that our group is strong when we share experiences and work together.

Changes in my household since joining the FFS/NAADS:

(i) Higher yields have led to increased household crop production – before FFS we often experienced periods of food shortage, my household is now food secure. (ii) All children in our household are now able to attend secondary school. (iii) Before, income was a gamble; now, income from agriculture is more secure and more income comes from non-agricultural activities (for example, FFS facilitation and training activities). (iv) We used to work as casual labour on other farmers' fields; today we occasionally hire other farmers to work for us. (v) My household has bought two cows from our increased income; we still have no oxen, but we are looking for one to buy; we now hire oxen for ploughing, using cash, instead of working for the owner of the oxen.

Moreover, in the sub-counties where FFS are present, a high proportion of NAADS group leaders, and parish and sub-county farmer fora members, were FFS graduates. As an example of the practical influence of the FFS philosophy on the functioning of NAADS groups and farmer fora, the chairs of Kyera sub-county farmer fora stated that FFS-turned-NAADS groups had been instrumental in assuring a rate of co-financing, as many of these groups had bank accounts from which they paid their NAADS fees.

Impact-assessment methodology

Impact assessments of agricultural research carried out by the CGIAR have in the past primarily focused on release of modern varieties and their associated economic returns from increased production (Pingali 2001; CGIAR 2004). Accountability is the predominant aim of these impact assessments, and their focus is chiefly on the interventions rather than on the communities in which these interventions were made.

The new Science Council (of 2004) broadened the scope of impact assessments, including assessments of 'soft' impact, such as training and capacity building (CGIAR 2004; Egelyng 2005). The use of participatory approaches in impact assessments has furthermore become more mainstream; participatory impact assessments take the target groups' perspective into account when testing the effects of an intervention.

Some researchers, however, have questioned whether using participatory approaches for impact assessment is the best way to obtain information about the intervention target group (Folke 2000). While participatory approaches in some situations are an effective way of extracting information, they have 'not in practice provided particularly good instruments for the kind of analysis of social relationships which projects require' (Mosse 1998:15). Further, participatory methods are 'often more likely to obscure than reveal the local social relations which shape them' (Mosse 1998:16).

Yet another approach to impact assessment, which has seldom been associated with the CGIAR, seeks to understand the dynamics and interactions between the intervention and the intervened (rather than the effects or impact of an intervention on the intervened). This approach, inspired by social-science development research, takes as its point of departure the dynamics of development and covers a wide spectrum of studies. While most impact-assessment studies see interventions as a one-way process, studies within this approach perceive impact as an interaction. Another characteristic of this development-research approach to impact assessment is that it is based on fieldwork and is rich in data.

This study is inspired by the development-research approach to impact assessment and aims to (1) place the agricultural technology development among poor farmers in Soroti district, Uganda in a socio-economic and institutional context, and (2) differentiate between different well-being categories when assessing the impact of access to improved technologies, farmer empowerment, and access to privatised demand-driven advisory services.

A team of researchers from the Danish Institute for International Studies (DIIS) and Agricultural Economics and Agribusiness at Makerere University carried out fieldwork in 2001 and twice in 2004. The fieldwork consisted of a range of complementary qualitative anthropological techniques with formal quantitative questionnaire surveys. The 2001 survey included interviews with local government officials, key informants among farmers, and farmer groups. It used qualitative Strengths, Weaknesses, Opportunities and Threats (SWOT) and Participatory Rural Appraisal (PRA) ranking techniques. The qualitative data collection was followed up with a formal questionnaire, employed with 106 randomly selected households. During the 2004 fieldwork, well-being indicators were identified, based on farmers' own perceptions. Using a district-based well-being ranking methodology (Ravnborg 1999; Ravnborg *et al.* 2004), a poverty index was constructed, based on 13 well-being indicators. A household survey was then developed on the basis of these indicators. The questionnaire was administered among 409 households, using stratified random sampling, including 307 households that were members of FFS and/or NAADS groups and 102 households that did not belong to any farmer group. Statistical analysis of quantitative data was followed up by qualitative in-depth life-history interviews with farmer-group members and leaders.

Characteristics of rural poverty in Soroti district

Well-being ranking methodology

Well-being indicators Multi-dimensional and participatory poverty well-being indicators were identified by farmers through a well-being ranking methodology developed and tested elsewhere in Uganda (Ravnborg 1999; Ravnborg *et al.* 2004; Boesen *et al.* 2004). Small groups of informants were asked to rank all households in their community into three groups, using a card-sorting method, and were then asked to describe the well-being of each group. The resulting sets of farmers' expressions of well-being were then extrapolated and tested statistically for representativeness within Soroti district and amalgamated into the 13 well-being indicators.[3]

Poverty index Based on the household-poverty indicators, a household poverty index was computed as the mean of its scores for each of the 13 well-being indicators. See Table 1. The variables of each well-being indicator are assigned the values (33), (67), and (100). Statistical analysis of the internal and external logic of the household poverty index was undertaken to confirm its validity.

The 25 and 70 percentiles, together with examination of the combined indicator score, provided guidance to defining the index values of three well-being categories: 'non-poor', poor, and very poor.[4] The actual values used in the poverty index are: 'non-poor' (<61.6), 'poor' (61.7–71.99), and 'very poor' (>72).

The well-being categories are heterogeneous. The categorisation of a given household is based on a calculation of the mean of all scores, rather than using the value of individual poverty indicators as determinants for a well-being category. A farmer without livestock (IANIMAL = 100) can, for example, still be characterised by the poverty index as 'non-poor', if the remaining 12 well-being indicators are sufficiently low. Similarly, a farmer who owns more than five acres of land (ILAND = 33) can be characterised as 'poor' if the mean score of all indicators is sufficiently low. This way of defining well-being categories allows for a dynamic analysis of the specific characteristics of poverty within a given enumeration area.

Characteristics of well-being in Soroti

With regard to assets, a significant correlation existed between ownership of land and well-being category. Approximately half of the farmers characterised as 'non-poor' by the poverty index owned more than five acres (2.0 ha), and the other half owned between one and five acres (0.4–2.0 ha); meanwhile, only a tenth of the 'very poor' had more than five acres, and close to a quarter had less than one acre. A similar significant correlation existed between well-being categories and ownership of animals, with three-quarters of the 'non-poor' owning cattle, while only 40 per cent of the 'poor' and 10 per cent of the 'very poor' did so. With regard to housing standards, the difference between the three well-being categories was less clear.

There was a significant correlation between well-being categories and non-agricultural income. Three-quarters of the 'non-poor' households received non-agricultural income, around half of which was from 'high-entry cost' sources, while none of the 'very poor' households received non-agricultural income from 'high-entry cost' sources; and about half of the 'very poor' had no non-agricultural income at all.

There were also significant correlations with agricultural labour. Less than a third of the 'non-poor' farmers worked for other farmers as casual labourers, and even then only to a limited extent. Meanwhile, some 90 per cent of the 'very poor' worked as casual labourers and most of them extensively. The opposite picture was the case for hiring casual labour: some 80 per cent of the 'non-poor' hired casual labour, while only 40 per cent of the 'poor' and 10 per cent of the 'very poor' did so.

Table 1: Scoring system for indicators constituting the household poverty index

Indicator	Score	Description
ILAND	33	Own (including leasehold, customary tenure, and freehold) more than 5 acres (2.0 ha) of land
	67	Own (including leasehold, customary tenure, and freehold) 1–5 acres (0.4–2.0 ha) of land
	100	Do not own land or own less than 1 acre (0.4 ha)
INONAG	33	Somebody from the household has 'high-entry cost' non-agricultural source of income, like being professional, having shop or business (trading, transport, etc.)
	67	Somebody from the household has non-agricultural source of income like tailoring, building, craft making, brewing beer, making and selling bricks, charcoal, or preparing and selling food
	100	Nobody from the household is engaged in non-agricultural sources of income
ILABOUR	33	Nobody from the household works for others as a casual labourer
	67	Somebody from the household works for others as a casual labourer, but either only 3 months or less per year, or more than 3 months per year but not more than once a week
	100	Somebody from the household works for others as a casual labourer more than 3 months per year, or less than 3 months per year but almost every day
IANIMAL	33	Somebody in the household has cattle or oxen, possibly together with other animals
	67	Nobody in the household has cattle, but they have other animals (goats, sheep, pigs, chickens, turkeys or rabbits)
	100	Nobody in the household has any animals
IHIRE	33	Hire labourers for at least two of the following tasks: land clearing, ploughing, planting, weeding, and harvesting
	67	Do not hire labourers or hire labourers for one task only
IFOOD	33	Have not experienced a period of food shortage within the last year
	67	Have experienced a period of food shortage within the last year which lasted less than 2 months or which lasted longer but the only recourse was eating less meat, using farm products rather than buying so much, or buying food
	100	Have experienced a period of food shortage within the last year which lasted 2 months or more
IFEED	33	Bought sugar when they last ran out of sugar, eat meat at least once a month, and fry food at least once a week
	67	Either did not buy sugar when they last ran out of sugar, or eat meat less than once a month, or fry food only occasionally (but not all three conditions at once)
	100	Went without sugar last time they ran out of sugar or rarely buy sugar, eat meat less than once a month *and* fry food only occasionally

(Table continued)

Table 1: Continued

Indicator	Score	Description
IHOUSING	33	Have houses with brick or plastered walls and iron or tile roofs
	67	Have houses which might have iron roof, plastered walls or walls of bricks or unburned bricks, but not both conditions at once
	100	Have houses with walls made of old tins or banana or other leaves, and grass-thatched roofs or roofs made of banana or other leaves, old tins or polythene, or have houses that are in need of major repairs
IHEALTH	67	Nobody in the household suffers from TB, HIV/AIDS, anaemia or chest-related disease, or is disabled
	100	Somebody in the household suffers from TB, HIV/AIDS, anaemia or chest-related disease, or is disabled
ISCHOOL	33	Have or have had children at secondary school or higher, or have children between 6 and 12 years in private or other schools at the same time as not having any children between 6 and 12 years who are not in school
	67	Have not (had) children in secondary school, and only have children between 6 and 12 years in public school while not having any children between 6 and 12 years who are not in school
	100	Have children between 6 and 12 years who are not in school
IDRESS	33	Woman owns shoes and both the woman and the children got new clothes within last 3 months
	67	Woman either does not own shoes or last got new clothes 6 months or more ago, or the children last got new clothes 6 months or more ago, or the woman does not own shoes and last got new clothes more than a year ago, but children last got new clothes in last 3 months
	100	Woman does not own shoes and both the woman and the children last got new clothes more than a year ago
IMARITAL	67	Household head is male or a married woman
	100	Household head is a widow or a single or divorced woman
IAGE	67	Either the household head or the wife is below 55 years of age
	100	Both the household head and the wife are at least 55 years old

Significant correlations also existed in terms of household food security and food consumption. Some 85 per cent of 'non-poor' households were food secure, compared with less than half of the 'poor' and less than a tenth of the 'very poor'. The differences between the well-being categories was less clear in terms of the type of food eaten (the indicator termed 'feeding' by the farmers).

Social well-being categories also showed significant differences between farmers from different well-being categories. More than a quarter of the 'very poor' had a household member who was seriously sick, compared with less than a tenth of the 'non-poor'. Almost a quarter of the 'poor' and 'very poor' households had children aged 6–12 years who did not attend school, while this was so for only one tenth of the 'non-poor'. Half the 'non-poor' households had children attending secondary or private schools, while only about a tenth of the 'poor' and 'very poor' did so. There were also clear differences in terms of marital status, with a third of the 'very poor' households being headed by a widow or a single or divorced woman, while this was so for less than a tenth of the 'non-poor' households.

Impact assessment of farmer empowerment, changes in opportunity structures, and access to improved technology

Processes of technology adaptation among members and non-members of farmer groups

Analysis reveals a close relationship between participation in farmer groups and effectiveness of agricultural production. A significantly higher percentage of farmers who were members of FFS/ NAADS groups adopted and used improved techniques for soil-erosion control, soil-fertility management, and pest management than did non-members (see Table 2).

There were significant differences between group members' and non-members' adoption of contour ploughing, grass-strip planting, and cover-crop planting. As for soil-fertility management, a significantly higher percentage of group members used improved techniques, including use of cattle manure, compost, or mineral fertilisers, while there was no significant difference in the use of traditional soil-fertility management techniques such as mulching. An even clearer picture emerged for pest management, where use of knowledge-demanding integrated pest management techniques was significantly higher among group members.

Technology development through FFS in Soroti district can be characterised as a group approach in which prototype technologies are adapted on group-managed plots through continuous monitoring of the crop and its growing conditions. Through this process, farmers' innovative capabilities to detect and solve field problems are enhanced. This form of agricultural development encourages and enables farmers to exchange ideas, experiment, and adapt technologies to specific local growing conditions, and organise and produce required local biological-based inputs, that is, botanicals (participatory observation).

Differentiation in well-being between members and non-members of farmer groups

Farmer-group membership is correlated with wealth at a 1 per cent level of significance (Table 3). Three times as many of the 'very poor' farmers were not group members as were group members.

Almost two-thirds of the farmers who were members of FFS/NAADS groups[5] were characterised as better off, compared with 41 per cent of non-group members. A particularly striking feature is that only 7 per cent of the FFS/NAADS group members were among the 'very poor' category of farmers (compared with 20 per cent of non-group members).

This situation can be interpreted as the result of two different processes. First, better-off farmers take advantage of their privileged social position in the local community to dominate groups that are associated with access to external resources. Second, poor farmers acquired skills through group membership that enabled them to escape poverty through improving the productivity of household resources. Depending on which of these processes dominates, the correlation can be interpreted as (1) a reflection of the formation of the group being biased towards better-off farmers, or (2) the effect of group membership contributing to poverty reduction. The study examined this question through a combination of qualitative interviews (focusing on the group-formation process and on the life-histories of well-off group members) and additional statistical analysis of questionnaire data (correlation and chi-squared tests between group membership and poverty indicators).

Table 2: Technology adaptation by members of NAADS and FFS groups in Soroti district

Soil erosion control	Members	Non-members
Contour ploughing ***	47%	43%
Planted grass strips ***	44%	46%
Planted cover crops **	18%	15%
Mulched [ns]	9%	1%
Made terraces [ns]	2%	7%
Fanya juu or *fanya chini* [ns]	5%	5%
Stopped removing plant residues [ns]	17%	22%
Soil fertility management		
Stopped burning [ns]	36%	36%
Use of green manure [ns]	21%	18%
Incorporated other residues ***	47%	42%
Used compost ***	24%	15%
Used chicken, goat or pig manure [ns]	37%	36%
Planted green manure ***	26%	14%
Used chemical fertiliser **	9%	4%
Used cattle manure to improve soil ***	37%	19%
Fallowed to improve soil ***	37%	29%
Mulched to improve soil fertility [ns]	2%	0%
Pest control		
Used improved seed***	47%	36%
Used natural enemy to destroy the pest***	29%	19%
Improved soil fertility ***	29%	16%
Monitored pest population **	59%	53%
Prepared the seed bed early enough **	48%	42%
Monitored weed population [ns]	45%	46%
Sprayed the crops ***	39%	27%
Did nothing to destroy the pests ***	0%	6%

Notes: $N = 409$ households. Figures are rounded up or down to nearest whole figure.
*** 0.01 level of significance; ** 0.05 level of significance;
ns = not significantly different.
Source: DIIS/MUK Soroti household survey 2004.

Group-formation process

The general lesson learned from group-formation processes elsewhere in Uganda and more widely in Africa is that when there are immediate tangible benefits to be gained from membership of a group, it is likely to be dominated by 'non-poor' farmers – if the group-formation process is 'open' and guided by local community leaders (Friis-Hansen 2003; Westerman *et al.* 2005).

Table 3: Social differentiation of NAADS and FFS groups in Soroti district, 2004

Poverty level ***	Member of a group (mostly FFS)		Overall
	Yes	**No**	
Better-off	63%	41%	57%
Less poor	30%	39%	32%
'Very poor'	7%	20%	10%

Note: *** 0.01 level of significance. N = 409 households. Figures are rounded up or down to nearest whole figure.
Source: DIIS/MUK Soroti household survey 2004.

Qualitative interviews with FFS members and local leaders revealed that because of successful 'sensitisation' prior to the group formation, the recruitment of members to FFS groups in Soroti district was done on the basis of self-selection (and exclusion) around a common 'interest in learning new skills'. As a result, the groups consisted of a mix of different types of farmer, indicating that the group-formation process was not biased towards better-off farmers.

In spite of the sensitisation, some farmers still joined FFS groups primarily because of an interest in accessing external funds. However, these farmers often left the group within a short period, when they realised that group activities were focused on experiential learning based on principles of informal adult education and did not provide its members with direct access to tangible goods. As a consequence, the FFS groups experienced a high initial membership turnover, with up to half leaving within the first year. These left in part because of faulty expectations (of direct material benefits), while better-off farmers left in part because they viewed participation as too time-consuming compared with the benefits. Common characteristics of farmers who remained are a willingness to invest time and effort in learning and in conducting joint activities. The majority of FFS members were women, and the FFS groups have been shown to be inclusive of illiterate and other socially disadvantaged people.

Using lifecycle interviews to assess change in well-being

In-depth lifecycle interviews with FFS/NAADS group members indicated that the majority of the better-off farmers had been among the 'poor' and 'very poor' categories when they joined FFS in 1999–2001. Women, many of whom were illiterate, formed the majority of the FFS/NAADS group members. The lifecycle interviews gave strong indications of reduced poverty among female members of FFS/NAADS groups.

On the basis of qualitative lifecycle interviews among FFS/NAADS group members, the following hypothesis was formulated:

> *Effective farmer learning in FFS, combined with improved technology access through NAADS, has created a pathway out of poverty based on improved agricultural productivity among poor farmers. Changes within the four agriculture-related poverty indicators are affecting the poverty indicator – whether somebody in the household works as a casual labourer (ILABOUR); whether the household is food-secure (IFOOD); whether the household hires casual labour (IHIRE); and whether income from surplus production has been invested in at least one cow (IANIMAL).*

Table 4: Membership of farmer groups by well-being indicators

Well-being indicator		— Membership of FFS/NAADS group					
		Yes			No		
		Count	%age of group members	Overall % within indicator	Count	Table N %	Column N %
	Total	307	75%		102	25%	
ILAND	33	119	29	39	38	9	37
	67	172	42	56	57	14	56
	100	16	4	5	7	2	7
INONAG	33	76	19	26	19	5	19
	67	125	32	43	56	14	57
	100	91	23	31	24	6	24
ILABOUR	33	167	**41**	**55**	38	9	37
	67	98	24	32	33	8	32
	100	40	10	13	31	8	30
IANIMAL	33	191	**47**	**62**	50	12	50
	67	102	25	33	37	9	37
	100	13	3	4	14	3	14
IHIRE	33	203	**50**	**67**	39	10	38
	67	100	25	33	63	16	62
IFOOD	33	206	**53**	**70**	48	12	49
	67	45	12	15	24	6	25
	100	42	11	14	26	7	27
IFEED	33	38	10	13	5	1	5
	67	223	58	78	81	21	84
	100	26	7	9	11	3	11
IHOUSING	33	10	3	3	5	1	5
	67	274	71	93	83	21	89
	100	10	3	3	5	1	5
IHEALTH	67	263	65	87	86	21	84
	100	40	10	13	17	4	17
ISCHOOL	33	90	31	39	16	5	26
	67	108	37	46	36	12	59
	100	36	12	15	9	3	15
IDRESS	33	118	30	39	30	8	31
	67	176	44	59	63	16	65
	100	7	2	2	4	1	4

(Table continued)

Table 4: Continued

Well-being indicator		Membership of FFS/NAADS group					
		Yes			No		
		Count	%age of group members	Overall % within indicator	Count	Table N %	Column N %
	Total	307		75%		102	25%
IMARITAL	33	269	66	87	86	21	84
	67	39	10	13	17	4	17
IAGE	33	274	69	91	93	23	93
	67	26	7	9	7	2	7

Note: The choice to use the mean instead assigning different value to the 13 well-being indicators is practical and not theoretically based. It would be very difficult to carry out a participatory prioritisation process of the indicators, because these are not known to the informants, as they are based on a consolidation process of statements about well-being. Figures are rounded up or down to nearest whole figure.

The first three of these four well-being indicators differ from the remaining 10 indicators in that they can change without the need for capital over long periods of time. Social mobility within the first three categories therefore seems plausible, compared with categories such as land ownership and the standard of housing, which are less likely to change within a period of four to five years. Quantitative statistics of household data are used to verify and test this hypothesis (below).

Significant correlation between 'being member of a group' and 'being better off'

To verify the hypothesis that there is a correlation between belonging to a group and the level of poverty within any of the 13 poverty indicators, we tested whether a correlation exists between membership of a group and being better-off within any of the 13 poverty indices. The results are shown in Table 4.

Some 307 households (75 per cent) belonged to a group, and 102 (25 per cent) did not. In terms of the first poverty index (ILAND), 109 households (28 per cent) were better off (meaning that the household owned more than five acres of land) *and* belonged to a group, while 39 per cent of people belonging to a group were better off.

Our hypothesis is statistically verified for the highlighted poverty indicators, ILABOUR, IANIMAL, IHIRE, and IFOOD (Table 4). In these four indicators, group membership is positively correlated with scoring 33 (that is, 'non-poor'). This suggests a strong relationship between being better off and being part of a group within these four areas.

In order to verify whether the numbers correlate, or that the large number of people being better off within the four poverty indices is just a reflection of statistical uncertainty, we carried out a chi-square test. The results show the following values: ILABOUR (0.000), IANIMAL (0.002), IHIRE (0.000), and IFOOD (0.001) – all *lower* than 0.05. This indicates that there is a significant correlation between these four indicators and relation to a group. And since a considerably larger number of better-off households within these poverty indices also belonged to a group (compared with the other indicators), it is valid to conclude that either being part of a group has a positive effect on poverty level or vice versa. Either way, the relation is significant (a significance value lower than 0.05 means that there is a minimum 95 per cent chance of being statistically correct when drawing this conclusion).

Conclusion

The central argument in this study is that a *combination* of farmer empowerment through experiential learning in FFS groups and changes in the opportunity structure through transformation of LG staff, establishment of sub-county farmer fora, and the emergence of private service providers has been successful in reducing rural poverty. Agricultural growth among poor farmers in Soroti district has been the key cause of poverty alleviation.

We have shown that farmers who were members of FFS/NAADS groups (at the time of the study) were significantly better off than non-members. The area-specific well-being ranking methodology used in this study (based on farmers' perceptions) proved a useful impact-assessment technique. Qualitative interviews further indicated that most farmers were among the 'poor' or 'very poor' when they joined an FFS. This is a major achievement and valuable evidence in support of the hypothesis that farmer empowerment through demand-driven advisory services can contribute significantly to alleviating rural poverty. The analysis further showed that pathways out of poverty included labour, food security, and investment in cattle.

The explanation for the higher rate of adoption of technology within FFS/NAADS groups is the *combination* of broad-based farmer learning in FFS with subsequent access to advice on commercial enterprises. A lesson learned is that the market-based spread of pro-poor technologies requires an institutional setting that combines farmer empowerment with an enabling policy environment.

Notes

1. The traditional Teso Farming System is based on the production of annual crops in a physical environment characterised by sandy infertile soils, heavy precipitation in two rainy seasons, and a prolonged dry season from December to March. The system is under considerable stress, resulting from severe cattle-rustling during the 1980s and 1990s; the prevalence of HIV/AIDS; insurgency and insecurity; increased variability of rainfall; and the collapse of the cotton marketing system. By the end of the 1990s rural households were experiencing widespread poverty and persistent food insecurity, which is an additional reason why many smallholder farmers are highly receptive to the opportunities offered by NAADS.
2. The concept of opportunity structures is further discussed in Friis-Hansen (2004b).
3. The result from the extrapolation analysis ensured that no major pattern of correlation existed between use and non-use of specific sets of indicators, making the result valid for all types of community and informant. Sets of indicators that were specific for local agro-ecological conditions (i.e. use of animal draught power) were left out of the final set of household poverty indicators. See Ravnborg *et al.* (2004) and Ravnborg (1999) for detailed discussion of the methodology.
4. The index values chosen for well-being categories were as follows: 'non-poor' have an index value below 61.6; 'less poor' consist of households with index value between 61.6 and 71.99; 'very poor' households have an index value of 72 and above. For additional technical discussions of computing a poverty index and selection of index values for well-being categories, see http://www.diis.dk/sw3086.asp.
5. The household survey was undertaken as a stratified random sample with a disproportional representation of FFS groups members/graduates. The majority of NAADS groups in this sample therefore comprise established FFS groups that registered as NAADS groups.

References

Boesen, J., R. Miiro, and S. Kasozi (2004) 'Basis for poverty reduction? A rich civil society, farmer innovation and agricultural service provision in Kabale, Uganda', *DCISM Working Paper* 2004:1, Copenhagen: Danish Institute for International Studies.

CGIAR (2004) 'Impact. Everybody's Business', *Annual Report 2003*, Washington, DC: CGIAR.

Duveskog, D. (2006) 'Theoretical Perspectives of the Learning Process in Farmer Field Schools', unpublished working paper, Nairobi: FAO.

Egelyng, H. (2005) 'Development Returns from Investing in International Agricultural Research, Vols. I and II', Policy Study presented at a seminar on agricultural research, Ministry of Foreign Affairs, Copenhagen, 4 March.

Folke, S. (2000) *Aid Impact: Development Interventions and Societal Processes*, Copenhagen: Centre for Development Research.

Friis-Hansen, E. (2003) 'Knowledge Management. Regional Thematic Review: Agricultural Technology Development and Transfer in Eastern and Southern Africa', *Report* No. 1347, Rome: IFAD.

Friis-Hansen, E. (2004a) 'Key Concepts and Experiences with Demand Driven Agricultural Extension in Tanzania. A Review of Recent Literature', *DIIS Working Paper* 2005:3, Copenhagen: Danish Institute for International Studies.

Friis-Hansen, E. (ed.) (2004b) *Farmer Empowerment. Experiences, Lessons Learned and Ways Forward. Volume 1: Technical Paper*, Copenhagen: Danida Technical Advisory Service, available at http://www.diis.dk/graphics/Events/2004/FarmerEmpowermentVol1Final.pdf (retrieved 5 September 2007).

IFAD (2002) *Regional Strategy Paper – Eastern and Southern Africa*, Rome: IFAD.

Mosse, D. (1998) *Development as a Process: Concepts and Methods for Working with Complexity*, London: Routledge.

Narayan, D. and P. Petesch (2002) *Voices of the Poor: From Many Lands*, Washington, DC and New York: Oxford University Press and World Bank.

Narayan, D., R. Chambers, M. K. Shah, and P. Petesch (2000a) *Voices of the Poor: Crying Out for Change*. Washington, DC: World Bank.

Narayan, D., R. Patel, K. Schafft, A. Rademacher and S. Koch-Schulte (2000b) *Voices of the Poor: Can Anyone Hear Us?* Washington, DC and New York: Oxford University Press and World Bank.

Parsons, D. J. (1970) 'Agricultural systems' in J. D. Jameson (ed.), *Agriculture in Uganda*, Oxford: Oxford University Press.

Pingali, P.L. (2001) *Milestones in Impact Assessment Research in the CGIAR 1970–1999*, Mexico DF: CGIAR Standing Panel on Impact Assessment.

Ravnborg, H.M. (1999) *Developing Regional Poverty Profiles Based on Local Perceptions*, Cali, Colombia: CIAT.

Ravnborg, H. M., J. Boesen and A. Sørensen (2004) 'Gendered district poverty profiles and poverty monitoring Kabarole, Masaka, Pallisa, Rakai and Tororo districts, Uganda', *DIIS Working Paper* 2004:1, Copenhagen: Danish Institute for International Studies.

Westerman, O., J. Ashby, and J. Pretty (2005) 'Gender and social capital: the importance of gender differences for the maturity and effectiveness of natural resource management group', *World Development* 33 (11): 1783–99.

The author

Esbern Friis-Hansen is an economic geographer with 25 years' experience as a development researcher and consultant, specialising in socio-economic and institutional issues in rural areas of East and Southern Africa. He is a senior research fellow in the Politics, Governance and Aid group at the Danish Institute for International Studies. Contact details: Danish Institute for International Studies, Strandgade 56, DK-1401K, Denmark. <efh@diis.dk>

No more adoption rates! Looking for empowerment in agricultural development programmes

Andrew Bartlett

The debate on empowerment encompasses an older discourse about the intrinsic value of empowerment, and a newer discourse about the instrumental benefits of empowerment; the concept of agency is useful in understanding this distinction. In agricultural development, empowerment efforts are often instrumentalist, viewed as an advanced form of participation that will improve project effectiveness, with adoption rates that promote compliance rather than intrinsic empowerment. Nevertheless, it is possible for projects to enhance the means for – and facilitate the process of – intrinsic empowerment. With regard to process, research and extension can make use of a constructivist rather than the behaviourist approach to support changes in knowledge, behaviour, and social relationships. In assessing empowerment, both developers and 'developees' need to look for evidence that people are taking control of their lives. Case studies – such as those used by the Indonesian Integrated Pest Management (IPM) Programme – will help to capture context and chronology, with unplanned behaviours being particularly useful indicators.

Power can be taken, but not given. The process of the taking is empowerment in itself.
(Gloria Steinem 1983)

... delightful as the pastime of measuring may be, it is the most futile of all occupations, and to submit to the decrees of the measurers the most servile of attitudes.
(Virginia Woolf 1929)

Understanding empowerment

A real-world example: IPM farmers in Nepal

Mr Ganga Ram Neupane is a rice farmer in Jhapa District of Nepal. In 1999, he attended a Farmer Field School (FFS), where he learned about integrated pest management (IPM). As a result, his yields increased by approximately 30 per cent, while costs remained the same; he spent more on compost and certified seed, but saved money by eliminating the use of pesticides.

The FFS attended by Mr Neupane was conducted by officials from the Plant Protection Department with support from the Food and Agriculture Organization of the United Nations (FAO). The benefits, in terms of increased yields and reduced use of pesticide, were anticipated in the project document. What is equally interesting, however, are the activities and the outcomes that were *not* planned (Bartlett 2002).

Mr Neupane discussed the benefits of IPM with other farmers. The Chairman of the Village Development Committee (VDC) decided to organise another field school, using VDC funds to pay for snacks, with Mr Neupane as a trainer. Mr Neupane subsequently organised FFS in a number of villages, travelling up to 20 km on his bicycle to conduct sessions.

Mrs Damanta Bimauli, another rice farmer in Jhapa, was not invited to an FFS, but she decided to 'gate-crash' the sessions in a neighbouring village. Back in her own village, she started a number of experiments to compare different types and rates of fertiliser, and she formed a group of farmers to discuss community issues. The group was interested in health, not just agriculture, and they played an active role in a polio-awareness campaign.

The Plant Protection Department has responded to these developments by organising workshops for 'farmer trainers'. This has led to further initiatives by IPM farmers. Mr Neupane and Mrs Bimauli are now members of the Jhapa IPM Association, an organisation that is run by farmers and which is linking up with similar associations in other districts. The association in Jhapa plans and organises its own training activities, and negotiates the support required from the Department of Agriculture and local government units.

A semantic dichotomy: instrumental versus intrinsic empowerment

Bertrand Russell (1936) wrote that power is the fundamental concept in social science 'in the same sense in which Energy is the fundamental concept in physics'. Discussion about how to acquire or share power has been at the centre of political debate for hundreds of years. It is only recently, however, that power has become part of the language of agricultural development.

Until the late 1990s, the term 'empowerment' was rarely heard among agricultural scientists, managers, and field workers. Now, it can be heard every day in project offices and training rooms across the world. Unfortunately, we are not always talking about the same thing.

The debate on empowerment encompasses two distinct discourses: an older discourse about the *intrinsic* value of empowerment, and a newer discourse about its *instrumental* benefits.

The intrinsic discourse has a long history, drawing examples from – and giving inspiration to – liberation movements in many parts of the world. The past 300 years have been filled with revolutions and civil wars, struggles for autonomy and greater rights. The empowerment of farmers has often been a part of these movements, seeking a combination of *Tierra y Libertad* – to quote a slogan of the Mexican Revolution (that is, access to land and the freedom to make independent decisions about what to produce), rather than being exploited by landlords, colonialists, corrupt governments and, more recently, multinational companies. In seventeenth-century England, Gerald Winstanley was leader of the revolutionary 'Diggers' and author of many pamphlets espousing their cause. He did not use the word 'empowerment', but that is clearly what he had in mind when he wrote:

> *True religion is this: to make restitution of the earth which hath been taken and held from the common people by the power of former conquests, and so set the oppressed free.*
> (Winstanley 1650)

Echoes of Winstanley's voice can still be heard. During the twentieth century, the literature on farmers' rights encompassed the work of hundreds of writers and activists from all parts of the globe, from Kropotkin and Gandhi, through Paulo Freire and Wendell Berry, to Vandana Shiva and Li Boguang.

The instrumental discourse on empowerment is much more recent. It was not until the second half of the twentieth century that management experts in industrialised countries started to recognise that job enrichment and worker participation could reduce costs and increase outputs. By the 1980s, Tom Peters and others were talking about 'empowering the workforce' (for example, Peters and Waterman 1982). This type of empowerment – as noted by Adrian Wilkinson (1997) – had little or nothing to do with older notions of industrial democracy and workers' rights, but had a lot to do with the effect that employee motivation has on efficiency and productivity. By the early 1990s, development agencies were applying the same rationale to women's empowerment, not as an end in itself, but as a contribution to other goals, as Naila Kabeer explains:

> *The persuasiveness of claims that women's empowerment has important policy payoffs in the field of fertility behaviour and demographic transition, children's welfare and infant mortality, economic growth and poverty alleviation has given rise to some unlikely advocates for women's empowerment . . . including the World Bank.* (Kabeer 2001: 17)

Following the Fourth World Conference on Women, held in Beijing in 1995, staff at the World Bank have broadened their interest in empowerment as a means for poverty reduction, applying the instrumental concept to a number of fields, including agricultural development. The organisation has generated a considerable volume of literature on the subject, giving characteristic attention to the issue of measurement (World Bank 2007).

With all this in mind, it is pertinent to ask what agricultural development professionals are talking about when they use the term 'empowerment'.

In agriculture, empowerment is often seen as the next step in the trend away from 'technology transfer' and towards increased participation, involving the diffusion not merely of information, but of expertise. As an advanced form of participation, empowerment entails farmers making their own decisions rather than adopting recommendations. It is expected that expert farmers will make 'better' decisions than outside experts, and that this will result in farming systems with a higher degree of productivity, efficiency, sustainability, and equity. In short, empowerment is a means for the achievement of goals that have been set by governments and donors. In this case we are talking about instrumental empowerment.

Rather than being the expression of any kind of liberation movement, the empowerment of farmers is often seen as a way of enhancing the effectiveness of projects and programmes that are planned and managed by the political and technical elite. This instrumental view of empowerment involves farmers taking greater control of livelihood assets in a way that is both predictable and non-threatening for other sections of society.

Notwithstanding the recent explosion of instrumentalist interest in the idea of farmer empowerment, there are plenty of people – researchers, educators, activists, and farmers themselves – who are talking about what they consider to be 'real' empowerment, that is, intrinsic empowerment. And there are others, many others, who are bewildered by the entire debate, suspecting that it is little more than the latest fashion in development rhetoric, but nevertheless making frequent use of the terminology in the hope that this will attract funding or the approval of their directors. It is not unusual to find, within a single development organisation or project, different employees who are espousing intrinsic, instrumental, and cynical views about the empowerment of farmers. The differences are not always recognised and are rarely analysed, with the result that it becomes difficult to plan, implement, and monitor activities that contribute to empowerment.

Conceptualisation: a transformational model of empowerment

Robert Chambers has succinctly defined empowerment in the context of rural development as follows: 'Empowerment means that people, especially poorer people, are enabled to take more control over their lives' (Chambers 1993: 12). Empowerment as it is understood by Chambers and Kabeer involves more than a few exceptional activities: instead it involves a profound and lasting change in the way people live their lives. In short, empowerment involves *a transformation.*

We can gain a better understanding of the nature of empowerment if we distinguish between three elements of this transformation: means, process, and ends (Bartlett 2004). *The means of empowerment* encompass a wide range of 'enabling factors', including rights, resources, capabilities, and opportunities. Means may be given or taken as part of the transformation; the key issue is what people do with those means. *The process of empowerment* is often seen in terms of 'making choices', but that is a simplification. The process involves a number of steps: analysis, decision making, and action. Only when the process is self-directed can we say that empowerment is taking place. *The ends of empowerment* are people taking greater control of their lives. In the case of rural development projects, this can be seen when certain social groups (for example, women, the poor, ethnic minorities) play a greater role in the management of livelihood assets, both in absolute terms and relative to other social groups.

As shown in Figure 1, all three elements of the transformation are needed for empowerment to take place. A change of *means*, on its own, may produce certain benefits such as access to services, but without *process* those benefits are a form of patronage, not empowerment. Conversely, attempts to change process without the means being in place will result in frustration and failure. Only when both means *and* process have been changed is it possible for the *ends* to be realised, and even then it may happen that the potential for empowerment is not converted into greater control, perhaps due to resistance from other social groups.

Generally speaking, a change in *means* creates the potential for a change in *process*. A change in *process* creates potential for a change in *ends*. In many cases, this transformation is cyclical, with a change in ends bringing about a further change in the means of empowerment.

The ends of empowerment can be described in absolute and/or relative terms, with the instrumental discourse often giving more attention to the former, while the intrinsic discourse usually gives more attention to the latter. For example, as a result of joining a savings group a woman may become absolutely empowered, because she can purchase goods and services that were previously out of her reach. She may also be relatively empowered as she takes on a larger role in family decision making, something that may have been previously dominated by her husband.

The essence of empowerment: it's all about 'agency'

If the debate about farmer empowerment is to become more meaningful, we need to foster a wider understanding of these concepts, particularly the difference between the instrumental

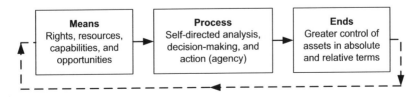

Figure 1: A transformation model of empowerment

and the intrinsic perspectives. What, then, is the difference? The answer rests in the distinction between farmers being *given* a greater role in *our* agenda, which we can call 'participation', and farmers *taking* control of their *own* agenda, which is what real empowerment is all about. The suggestion that participation is part of the instrumentalist agenda may be an unfair simplification of the wide range of opinions and methods that are associated with participatory approaches in rural development, as delineated by writers such as Nour-Eddine Sellamna (1999), but it helps to clarify the distinction.

At the root of the distinction is the concept of 'agency', which most social scientists recognise as a crucial component of empowerment, but which is not found in some forms of participation. The concept of agency stems from the idea of the 'human agent'. Rural people can become the agents of their own development, or they can remain the objects of somebody else's development process. As explained by Paulo Freire more than 30 years ago, farmers become 'agents of change' through purposive action which effects and demonstrates greater control over their lives:

> If a social worker (in the broadest sense) supposes that s/he is 'the agent of change', it is with difficulty that s/he will see the obvious fact that, if the task is to be really educational and liberating, those with whom s/he works cannot be the objects of her actions. Rather they too will be agents of change. If social workers cannot perceive this, they will succeed only in manipulating, steering and 'domesticating.' (Freire 1976: 114)

Agency involves a self-directed process, not only in the practical sense of a person carrying out activities that impinge upon the material world, but also in a deeper ontological sense that involves the construction of that person and their world. When empowerment occurs, this deeper process manifests itself in lasting changes in perceptions and relationships. Recognition of this transformation is essential to the intrinsic – rather than instrumental – view of empowerment, as becomes clear in this eloquent quote from Naila Kabeer:

> Agency is about more than observable action; it also encompasses the meaning, motivation and purpose which individuals bring to their activity, their sense of agency, or 'the power within'. While agency often tends to be operationalised as 'individual decision making', particularly in the mainstream economic literature, in reality, it encompasses a much wider range of purposive actions, including bargaining, negotiation, deception, manipulation, subversion, resistance and protest as well as the more intangible, cognitive processes of reflection and analysis. (Kabeer 2001: 21)

The instrumental view of empowerment is problematic because it involves the intention to both promote *and* constrain agency. It involves giving rural people greater choice within predetermined boundaries. It welcomes negotiation, but shuns resistance. For example, farmers are allowed to manage their own community-based organisations in accordance with regulations provided by the donor or the state; they are given an opportunity to participate in the development and testing of technology that subsequently requires official approval before it can be widely used; they are taught to make their own decisions about crop management at the same time as being put under pressure to adopt or reject certain practices. In all of these examples, agency is *localised*, it is limited to decisions and action taken within narrow technical and/or social parameters. This is participation.

Increased participation can bring considerable benefits to rural people, and it can lay the foundations for genuine self-determination, but until agency becomes *generalised*, until relationships begin to change and the consequences of this change become unpredictable, it does not merit the use of the term 'empowerment'. Consequently, the remainder of this article will limit the use of this term to the personal and social transformation that is envisaged in the intrinsic discourse.

Corollary: choosing between agency and adoption

If agency is the key to real empowerment, then empowerment is not something that we – as policy makers, agricultural scientists, and development workers – can do to rural people; rather it is a consequence of something that rural people do for themselves. Although agricultural development projects can be implemented in ways that initiate and support the empowerment of farmers, empowerment itself will always be outside the project framework. Empowerment cannot be seen as a sequence of project activities, nor can it be reduced to a measurable objective; instead, it involves rural people setting their own goals, managing their own activities, and assessing their own performance.

Adoption rates are the antithesis of agency in agricultural development projects. In itself, the adoption process provides scope for reflection and analysis by farmers. But by using adoption rates in the design and evaluation of projects, development professionals are promoting compliance rather than real empowerment. These adoption rates take a variety of forms, for example: 'improved varieties will be used on 10,000 hectares', 'at least 25 per cent of households will have ceased shifting cultivation and planted fruit trees', 'average pesticide applications will be cut by a half'. In each case, project planners and managers have decided how rural people should live their lives. Planners and managers usually do this with the best of intentions, but they remain in control of the development process. They are the 'developers'.

Despite all the talk of partnerships, most development projects in the agriculture sector continue to be characterised by a sharp distinction between the developers and the 'developees'. Every few years, we find a new label for the developees in an attempt to demonstrate greater political correctness. The past 25 years have seen a shift in usage from 'audience', 'recipients', and 'beneficiaries' towards 'actors', 'participants', and 'collaborators'. But in every case, the instigators of development, the people who fund and administer and evaluate projects, are talking about *somebody else*.

If agricultural projects were to focus on agency rather than adoption, the distinction between developer and developee would begin to fall apart. As a corollary of empowerment, we – as development professionals – must also become subjects of the development process. If we want farmers to gain power, we must expect to lose some ourselves. It is unrealistic to promote changes in the relationships of rural people – between the poor and the rich, between women and men – without also being open to changes in our own relationship with the people we set out to help. We cannot exempt ourselves from the transformation that we claim to support.

Implications for practice: seeking entry points for empowerment

What are the implications of this transformation for the design of agricultural development programmes? There is no single answer; rather there is a spectrum of possibilities. At one extreme, the post-development critique of Arturo Escobar (1994) suggests that real empowerment cannot be a sub-set of development; self-determination is not something that can be managed and measured by development professionals. Indeed, development and professionalism are parts of the problem, not the solution, and real empowerment will be a consequence of social movements, not the result of projects supported by the World Bank, FAO, and IRRI (International Rice Research Institute).

At the other end of the spectrum, there have been attempts to use agricultural interventions as 'entry points' for empowerment. Instead of trying to make empowerment processes a sub-set of development, development activities become a sub-set of empowerment. This entails programmes that are carefully planned and organised at the outset, but which become increasingly flexible and open-ended as farmers start to demonstrate greater agency. The IPM Farmer Field School has been used in this way, as illustrated in the opening section of this article and explained further by John Pontius *et al.* (2002).

If we accept that development projects can play a role in creating entry points for empowerment, the transformational model suggests that this can be achieved in two different ways: by enhancing the means of empowerment, and by facilitating the process.

There is a wide range of possibilities for enhancing the means of empowerment, including land reform, the provision of credit, and the regulation of markets. These interventions can have an indirect effect on empowerment by expanding the opportunities that farmers face, and by reducing the risk associated with making choices.

By contrast, research and extension organisations have often attempted to have a more direct impact on the choices that farmers make. New technology can be seen as an opportunity, but it has frequently been presented as a prescription. Extension programmes have tried to bring about direct changes in three domains – knowledge, behaviour, and social relations – by means of interventions that are based on a behaviourist model of learning and a transmission model of communication (see Figure 2).

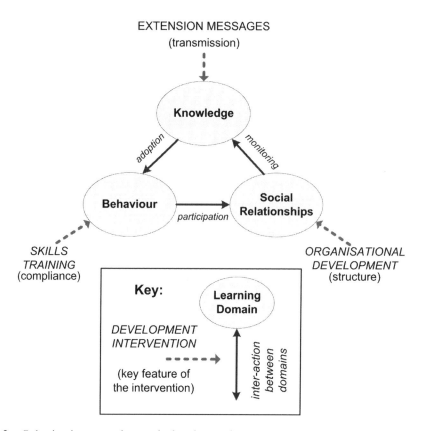

Figure 2: Behaviourist approach to agricultural extension

As part of the behaviourist model, the mind of the learner is treated as a 'black box' which responds to external stimuli. What happens inside the box is considered to be irrelevant. By providing the learner with appropriate information, and giving reinforcement through punishments and rewards, desired patterns of behaviour can be produced in a predictable and measurable manner.

The behaviourist approach is a hindrance to empowerment, because it does not recognise the importance of agency. The alternative is a constructivist approach to learning, which assumes that knowledge, behaviour, and social relations cannot be transmitted from one party to another, but must be uniquely created by the human agent as a consequence of critical thinking, experimentation, and communicative action (see Figure 3).

In practice, the constructivist approach requires interventions that foster agency during the interaction between the three domains – in other words, interventions that facilitate the process of empowerment. For example, at the point of interaction between knowledge and behaviour, agency can be stimulated through experiential learning – by encouraging and supporting a process of observation and experimentation, farmers generate *their own* knowledge and make *their own* decisions about behaviour. This learning process has a number of variants that have been given different names: emancipatory, transformational, discovery-based. The critical feature of all of these variants is *ownership* by the learner, not just of the outcomes but, increasingly, of the process itself.

> *Like power, knowledge is not something that is possessed, accumulated and unproblematically imposed upon others. Nor can it be measured precisely in terms of quantity or quality. It emerges out of processes of social interaction.* (Long and Villareal 1993: 157)

Behaviourist and constructivist approaches are not mutually exclusive. It is possible – and often desirable – to deliver information to farmers *and* facilitate experiential learning, to conduct skills training *and* facilitate communicative action. What we need to consider are the links between the two approaches: which takes precedence? Are we delivering information to farmers so that they have greater opportunities for experiential learning, or are we facilitating

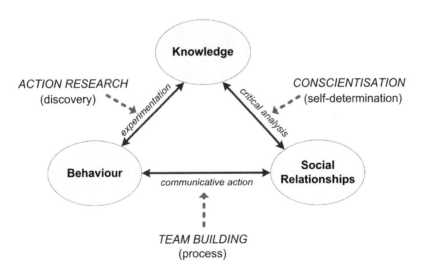

Figure 3: Constructivist approach to agricultural extension

experiential learning so that farmers know what to do with the information we are delivering? Are projects as a whole designed to maintain current relationships or transform them?

Assessing empowerment

Strategies: keeping agency in the assessment relationship

If agency is the key to real empowerment, it should also be the key to the assessment of projects that aim to promote empowerment. Impact assessment becomes a search for consequent agency. Donors and implementing organisations cannot decide the precise outcomes of empowerment for rural people; farmers have to do that for themselves. What planners and managers need to do is look for evidence that farmers are taking control of their lives, and determine how helpful project interventions have been in this process.

Where are we going to find agency, and how are we going to find it? To answer this question it is useful to recognise that assessment is the manifestation of a relationship within or between groups of people.

Table 1 shows a typology of five relationships and the assessment strategies that could be used to examine agency in each. The first two relationships in this typology (A and B) have been given the name 'empowerment evaluation' by David Fetterman *et al.* (1996), because they are 'designed to help people help themselves'. In both cases, assessment involves reflection and analysis by the developee, with performance and impact measured against indicators that he or she has selected. The difference between the two types of assessment is that participatory monitoring and evaluation (PM&E) is inward-looking, while the accountability mechanism is outward-looking. PM&E could, for example, involve members of a farmers' group setting their own targets and monitoring their own performance. As part of an accountability mechanism, however, the group could set indicators for the services to which they are entitled, and monitor the performance of the local government or a development project. In the first example they would present the results to themselves, but in the second example they might present their findings in a public forum.

A wide range of PM&E tools and techniques has been developed in the context of agricultural development projects, but they rarely challenge the power relationship between the developer and the developee. By comparison, far less work has been done to develop accountability mechanisms, perhaps because accountability is – in itself – a relationship of power between two actors: the 'object' of accountability, the one obliged to account for his/her actions, and the 'agent' of accountability, the one entitled to demand answers. A more detailed analysis of this relationship has been carried out in a number of works by Anne-Marie Goetz and Rob Jenkins (for example, Goetz and Jenkins 2001).

Table 1:   A typology of assessment options for development projects that aim to promote empowerment

	Assessment relationship	**Assessment strategy**
A	Self-assessment by developee	Participatory monitoring and evaluation
B	Developee assesses developer	Accountability mechanisms
C	Negotiated assessment by both	Constructivist evaluation (fourth generation)
D	Developer assesses developee	Applied anthropology
E	Self-assessment by developer	Action research

From an epistemological point of view, the third relationship in the typology, the negotiated assessment (C), is considerably more complicated than the other four, because it involves an attempt to explore and reconcile the perceptions and experiences of different groups of stakeholders. This type of assessment has been given the name 'Fourth Generation Evaluation' by Egon Guba and Yvonne Lincoln (1989), the authors of the basic text on the subject, who describe their methodology as 'hermeneutic-dialecticism'. Whatever you call it, negotiated assessment involves an encounter between agents who may have very different worldviews (that is, constructions), and it requires considerable skill – usually from an outside facilitator – if the views of all agents are to be treated as equally valid.

The next relationship in the typology, the assessment of the developee by the developer (D), is perhaps the most common type of assessment in agricultural development programmes. If we are looking for evidence of agency, however, the developer should focus on the processes undertaken by the developee, not only on the outcomes of those processes. This is not process evaluation as normally understood (that is, as a type of formative evaluation that focuses on how interventions are conducted); rather it is a form of applied anthropology, involving observations of interactions among stakeholders and, possibly, surveys of their opinions. The use of anthropological methods does not mean that the analysis is free of any judgements. It is possible to have as our goal the existence of a process with certain characteristics and, consequently, to examine the impact of our interventions on the occurrence of that process. Trying to assess human agency as a single process is enormously difficult, but we can break it into subprocesses that are easier to observe and record, both qualitatively and quantitatively. Drawing on the 'learning domains' described above, we can look for agency in *how* stakeholder knowledge, behaviour, and social relations change over time; more specifically, we would examine the type and incidence of experimentation, communicative action, and critical analysis.

Action research is the fifth and last type of assessment in the typology. This is, of course, similar to the PM&E undertaken by farmers or other developees. Nevertheless, there is a distinct literature describing the methods that professionals can use for self-assessment. The attention given to agency is perhaps most intense in *emancipatory* or *critical* action research. According to Ortrun Zuber-Skerritt (1996: 4), this involves 'participants' emancipation from the dictates of tradition, self-deception and coercion; their critique of bureaucratic systematisation [and] transformation of the organisation and of the educational system'. This sounds a lot like empowerment, not of the developees, however – but of the developers themselves.

This typology is not exhaustive. It would be possible to add assessment relationships between two groups of developees (for example, a farmer exchange) and between two groups of developers (for example, a peer review). What has been excluded deliberately is any kind of *independent* evaluation, involving an attempt by outsiders to assess what the developers and/or the developees have achieved. An independent evaluation precludes the need for agency by either party in the development relationship. Indeed, such an evaluation, by attempting to assign an official value to decisions and actions taken by the developer or developee, would be inherently disempowering. If outsiders are involved in the assessment of empowerment, the most appropriate roles are facilitator or resource person, not judge and jury.

Indicators: focusing attention on the behaviour of stakeholders

For each of the above strategies, the stakeholders in the assessment should be able to identify a range of indicators. The literature on assessing empowerment is full of suggestions for indicators, many of which focus on behaviours that demonstrate agency (for example, Malhotra *et al.* 2002; Oakley 2001). A checklist of behavioural indicators developed by one project in Bangladesh is shown in Table 2.

Table 2: Behavioural indicators used by field staff of a Rural Livelihoods Programme in Bangladesh (Bartlett 2004)

Behaviour category	Specific indicators
Organisational behaviour	• Women in leadership roles • Active participation in group decision making • Self-determined collective action
Planning behaviour	• Setting own goals • Agreeing upon and implementing a strategy towards the achievement of goals • Self-monitoring of progress and achievements
Entitlement behaviour	• Exercising rights • Making claims as individuals or groups • Engaging in advocacy
Economic behaviour	• Holding and using cash • Making sales, purchases, leases • Negotiating wage rates
Learning behaviour	• Seeking information • Taking action to share knowledge with others
Experimental behaviour	• Testing and modifying technologies • Rejecting a recommended technology as a result of critical thinking

While the categories of behaviour shown in Table 2 could be useful elsewhere, the selection of precise indicators must be done on a case-by-case basis, not least because certain behaviours can mean different things in different situations. Behaviours such as carrying out field trials, making use of credit, or joining community organisations could indicate new-found self-determination for certain people in particular situations, but could be considered routine for other people, or occur as a response to coercion in other situations.

Methods: using case studies to contextualise empowerment

In view of the contextual nature of empowerment, case studies are a useful assessment method that can be carried out as part of all five assessment strategies described above. When undertaken in a rigorous manner, case studies can capture both process and outcomes, using a combination of qualitative and quantitative indicators, and involving alternative ways of collecting and presenting information (for example, interviews, data sheets, photographs, maps).

As a methodology for impact assessment, case studies have strengths and weaknesses. Among the strengths is the opportunity for exploring the chronology that is inherent in any process; by providing information about the 'before and after' situations, plus an examination of the sequence of events that connected the two situations, a case study can go some way to establishing a counterfactual and attributing certain changes to particular interventions.

Among the weaknesses is the 'microscopic' nature of the case study, which reduces the possibility of making generalisations. The validity of case-study research can be improved, however, by using multi-method cases that triangulate between different types of data, and by carrying out multiple studies with cross-case analyses. Robert Yin (2002) describes a number of procedures for strengthening case studies, including techniques such as pattern matching.

Questions have also been raised about the reliability of case studies, because they often make use of qualitative data and subjective assessments. Rather than being a drawback, these characteristics may be helpful in the assessment of empowerment. The personal feelings and interpretations of stakeholders can be used as indicators of agency, particularly if there is an evident connection between changes in perception and changes in behaviour. Nevertheless, steps can be taken to improve the reliability of case studies by reducing the subjectivity of investigators. This can be done by establishing a rigorous protocol for the collection of case data, and by carrying out a peer review of the completed studies.

Back to the real world: assessing farmer empowerment in Indonesia

In 1997 and 1998, field staff of the Indonesian National Integrated Pest Management (IPM) Programme produced a large set of case studies using a combination of quantitative and qualitative methods. The purpose of each study was 'to present a description and analysis of the development achieved by IPM trained farmers in one sub-district' (Susianto *et al.* 1998: 1).

Case studies were produced for 182 sub-districts, each consisting of maps, chronologies, quotations, photographs, economic analysis, and various tables. In total, more than 3000 pages of information were compiled over a six-month period. Three particular processes were examined in the studies:

- farmer field schools (FFS) and associated interventions organised under the National IPM Programme;
- farmer-to-farmer activities that were planned and organised by FFS alumni with minimal outside support;
- changes in relationships among farmers, and between farmers and the government, referred to as 'social gains'.

Figure 4 shows how these processes can be related to the constructivist learning.

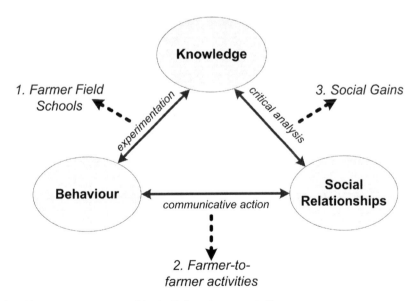

Figure 4: Three processes assessed in the Indonesian case studies

The farmers and field staff who carried out these studies attended methodological workshops, where they were provided with an outline of the issues to be covered in the cases. Using these outlines as a basis, they developed their own indicators and methods. Subsequently, meetings were held to discuss the information that had been collected and to review drafts of the study reports.

The process of producing these case studies, and the final results, were useful at three different levels: in the selected villages, where farmers ('developees') were able to participate in self-assessment; at the sub-district level, where teams of field staff ('developers') were also involved in self-evaluation; and at the national level, where the cases were examined for patterns and exceptions. At each level, the cases made an important contribution to team building and management decision making.

Independent cross-case analyses were carried out at a later date by Henk van den Berg (2004), during which two types of impact data were extracted from the case studies: (a) the incidence of 'spontaneous behaviour', as an indicator of empowerment; and (b) trends in pesticide sales, as an indicator of economic impact.

In the first of these cross-case analyses, 62 types of behaviour were identified, which, having been observed although not planned, could be reasonably attributed to what farmers had learned through the IPM programme. The frequency of each type of behaviour was determined by examining the 182 case studies. For example, farmers conducted their own field studies in 180 sub-districts (98.9 per cent); farmer trainer associations were formed in 35 sub-districts (19.2 per cent); pesticides were removed from village credit packages due to farmers' protests in 33 sub-districts (18.1 per cent); and farmer-to-farmer sales of alternative inputs occurred in 14 sub-districts (17.7 per cent). These figures are *not* adoption rates as normally understood, because the list of behaviours was drawn up *ex post*, not *ex ante*. What *is* being measured is the breadth and depth of self-determination among IPM farmers – in other words, *agency*.

The author of the cross-case analyses concluded that 'Substantial and widespread evidence from Indonesia suggests that FFS related project activities provide an impetus for spontaneous local programs with multiple impacts. The diversity of activities is indicative of farmer creativity and situational differences' (van den Berg 2004: 25). In addition, the data on pesticide sales in eight sub-districts showed a decline of between 70 per cent and 99 per cent, leading to the conclusion that there was 'a clear association between strong local IPM programmes and a drastic reduction in pesticide sales' (van den Berg 2004: 26).

It is worth noting that the cross-case analyses, drawn from 182 case studies in Indonesia, were part of a broader synthesis of 25 impact studies which had been conducted in 11 countries over a ten-year period. The methodology of the 25 studies was highly diverse, but pattern matching during the meta-analysis allowed valid conclusions to be drawn about the general benefits of IPM farmer field schools.

Impact assessment, like any other activity, has costs and benefits. Taking the Indonesian example given above, it may seem that the multiple case-study method was a hugely expensive way of collecting the data needed to assess empowerment and changes in pesticide use. Such a conclusion would be mistaken. The case studies were primarily designed to (a) facilitate the empowerment process by providing opportunities for communicative action and critical analysis, and (b) strengthen the management of the IPM programme, both within and outside the government apparatus. The generation of data that could be used for *ex-post* impact assessment was a side effect, a spin-off of these multifaceted processes.

The fact is that impact assessment often happens like this, as an after-thought that becomes a scrabble for data. In the Indonesian IPM programme, the data were available in large quantities, generated not by the monitoring and evaluation unit of a donor agency or government department, but by farmers and field staff who had been engaged in empowering processes over a period of more than a decade.

Final observations

Promoting real empowerment through agricultural development programmes is like raising fish in rice fields – it can be done, but the conditions are not ideal.

What is the problem? Quite simply, development projects are inherently instrumental. Projects are designed to achieve specific objectives by means of a predetermined sequence of activities that generate quantifiable outputs. Planned levels of outputs are used as indicators of success, and may be a key part of the contractual agreement between donor, government, and implementing organisation. Consequently, the opportunities for self-determination among the stakeholders are limited from the outset.

Within the limitations set by project objectives, there is usually scope for different approaches to be used, some of which are more empowering than others. But most agricultural development projects do not prioritise empowerment. Instead, these projects are designed to promote poverty alleviation or food security or the conservation of natural resources. These are important goals, but there is a widespread assumption that they can best be achieved through a behaviourist approach. Consequently, rural people become the object of production targets, approved varieties, recommended practices, demonstration plots, model farmers, and *adoption rates*, all of which are contrary to constructivist learning and the expression of agency.

In his review of agricultural development schemes that took place across the world during the mid and late twentieth century, James Scott (1998: 286) wrote that the 'unspoken logic behind most of the state projects for agricultural modernization was one of consolidating the power of central institutions and diminishing the autonomy of cultivators and their communities'. A decade later, the agenda of modernisation, rooted in positivist science and implemented by means of behaviourist projects, continues to pervade much of agricultural development. It is hard to reconcile this agenda with the transformational nature of real empowerment.

But occasionally it can be done, as the IPM farmers in Nepal and Indonesia have demonstrated. They have taken the opportunities provided by agricultural projects and turned them into something else – a process in which the distinction between developers and developees has become blurred, and in which unexpected outcomes may be a better indicator of success than planned outputs.

References

Bartlett, Andrew (2002) 'Farmer in Action: How IPM Training is Transforming the Role of Farmers in Nepal's Agricultural Development', Kathmandu: FAO.

Bartlett, Andrew (2004) 'Entry Points for Empowerment', Dhaka: CARE International, available at http://communityipm.org/docs/Bartlett-EntryPoints-20Jun04.pdf (retrieved 12 August 2007).

Chambers, Robert (1993) *Challenging the Professions: Frontiers for Rural Development*, London: ITDG.

Escobar, Arturo (1994) *Encountering Development: The Making and Unmaking of the Third World*, Princeton, NJ: Princeton University Press.

Fetterman, David, Shakeh Kaftarian, and Abraham Wandersman (1996) *Empowerment Evaluation: Knowledge and Tools for Self-Assessment and Accountability*, London: Sage.

Freire, Paulo (1976) *Education: The Practice of Freedom*, London: Writers and Readers Publishing Cooperative.

Goetz, Anne-Marie and Rob Jenkins (2001) *Voice, Accountability and Human Development: The Emergence of a New Agenda*, New York, NY: UNDP.

Guba, Egon and Yvonne Lincoln (1989) *Fourth Generation Evaluation*, London: Sage.

Kabeer, Naila (2001) *Discussing Women's Empowerment: Theory and Practice*, SIDA Studies No. 3, Stockholm: Swedish International Development Agency (SIDA).

Long, Norman and Magdalena Villareal (1993) 'Exploring development interfaces: from transfer of knowledge to the transformation of meaning', in F. J. Schuurman (ed.) *Beyond the Impasse: New Directions in Development Theory*, London: Zed Books.

Malhotra, Anju, Sidney Ruth Schuler, and Carol Boender (2002) 'Measuring Women's Empowerment as a Variable in International Development', Washington, DC: World Bank.

Oakley, Peter (ed.) (2001) *Evaluating Empowerment: Reviewing the Concepts and Practice*, Oxford: INTRAC.

Peters, Tom and Robert Waterman (1982) *In Search of Excellence: Lessons from America's Best-run Companies*, New York, NY: Harper & Row.

Pontius, John, Russ Dilts, and Andrew Bartlett (2002) *From Farmer Field School to Community IPM: Ten Years of IPM Training in Asia*, Bangkok: FAO, available at http://www.fao.org/docrep/005/ac834e/ac834e00.htm (retrieved 12 August 2007).

Russell, Bertrand (1936) *Power: A New Social Analysis*, London: George Allen & Unwin.

Scott, James C. (1998) *Seeing Like a State: How Certain Schemes to Improve the Human Condition Have Failed*, Yale, CT: Yale University Press.

Sellamna, Nour-Eddine (1999) 'Relativism in agricultural research and development: is participation a post-modern concept?', *ODI Working Paper* 119, London: Overseas Development Institute (ODI).

Steinem, Gloria (1983) 'Far from the opposite shore', in *Outrageous Acts and Everyday Rebellions*, New York: Holt, Rinehart and Winston.

Susianto, Agus, Didk Puwadi, and John Pontius (1998) 'Kaligondang: A Case History of an IPM Sub-district', Jakarta: FAO, available at http://www.communityipm.org/docs/Kaligondang.zip (retrieved 12 August 2007).

van den Berg, Henk (2004) 'IPM Farmer Field Schools: A Synthesis of 25 Impact Evaluations', Rome: FAO, available at http://www.fao.org/docrep/006/ad487e/ad487e00.htm (retrieved 25 August 2007).

Wilkinson, Adrian (1997) 'Empowerment: theory and practice', *Personnel Review* 27 (1): 40–56.

Winstanley, Gerald (1650) 'A New-Yeers Gift for the Parliament and the Armie', quoted in Kenneth Rexroth 'Communalism: From its Origins to the Twentieth Century' available at http://www.diggers.org/rexroth_diggers.htm (retrieved 12 August 2007).

Woolf, Virginia (1929) *A Room of One's Own*, London: Hogarth Press.

World Bank (2007) 'PovertyNet – Empowerment', internet portal, available at http://go.worldbank.org/S9B3DNEZ00 (retrieved 12 August 2007).

Yin, Robert K. (2002) *Case Study Research: Design and Methods* (3rd edn), London: Sage.

Zuber-Skerritt, Ortrun (1996) *New Directions in Action Research*, London: Falmer Press.

Appraisal of methods to evaluate farmer field schools

Francesca Mancini and Janice Jiggins

The need to increase agricultural sustainability has induced the government of India to promote the adoption of integrated pest management (IPM). An evaluation of cotton-based conventional and IPM farming systems was conducted in India (2002–2004). The farmers managing the IPM farms had participated in discovery-based ecological training, namely Farmer Field Schools (FFS). The evaluation included five impact areas: (1) the ecological footprint and (2) occupational hazard of cotton production; and the effects of IPM adoption on (3) labour allocation; (4) management practices; and (5) livelihoods. The analysis showed that a mix of approaches increased the depth and the relevance of the findings. Participatory and conventional methods were complementary. The study also revealed different impacts on the livelihoods of women and men, and wealthy and poor farmers, and demonstrated that the value of the experience can be captured also in terms of the farmers' own frames of reference. The evaluation process consumed considerable resources, indicating that proper budgetary allocations need to be made.

Introduction

The use of participatory methods to enhance the effectiveness of research and technology development in agriculture has found increasing support from institutions and donors since the 1980s. Declining productivity and increasing production costs are affecting the economic sustainability of small-scale farming in many developing countries, thereby compromising an important source of income for millions of people. This critical situation is particularly evident in the Indian cotton-farming sector, largely because of the massive and ineffective use of pesticides (Shetty 2003). In order to reverse these negative trends, the Indian government has been supporting innovative extension systems, which conceptualise the diffusion of technology as a social process based on adaptation and innovation by the users.

A notable example of participatory research and extension that engages researchers, extensionists, and farmers in on-site experimentation is the Farmer Field School (FFS). FFSs were conceptualised in the 1970s and 1980s and first implemented in Indonesia in 1989 to deal with a widespread outbreak of a rice pest that threatened Indonesia's food security (Pontius *et al.* 2002). A few pioneering entomologists demonstrated that it was the enormous use of pesticides, which had been promoted by the government to control brown plant hoppers, that was the primary cause of the outbreak. Because agro-ecological relationships are inherently place-dependent and time-specific, it is ineffective to make decisions on the basis of the universal dissemination of standard technologies and simple messages in order to control crop pests. FFSs were therefore organised to teach rice farmers how to observe and measure, in their own fields, the ecological relationships underlying integrated pest management (IPM), and thereby help them to reduce their reliance on chemical pest controls (Kenmore 1996).

The focus of the FFS was, and still is, on learning through discovery, experimentation, informed decision making, and group or community leadership and action. FFSs thus have social goals beyond mere changes in pest-management techniques: goals that seek to position farmers as field experts, who collaborate with the extension staff to find solutions relevant to the local realities. FFS programmes emphasise farmers' ownership of development processes, partnership with other development agents, and group collaboration. Evaluations of the accomplishments of various FFS programmes agree in their main conclusion that attending an FFS strengthens farmers' ecological knowledge of pests and predators. In most reported cases, the understanding of the crop ecosystems has induced a reduction in pesticide use, as well as higher yields and profits – for instance, in sweet potato, potato, and cotton production systems. So far, the long-term sustainability of the changes in pesticide use has been questioned by only two studies, one on vegetables and one on rice, which have drawn opposite conclusions (Khalid 2002; Feder *et al.* 2004). Some authors argue that the economic feasibility of FFS programmes is in question, because it depends on the diffusion rate of the outcomes from the participants to neighbouring farmers. Studies conducted to investigate this diffusion have led to contradictory findings, perhaps because the diffusion of knowledge depends strictly on context and content (for example, Rola *et al.* 2002). FFS supporters shift the focus of the economic debate to the benefits of IPM-FFSs, contending that a reduction in pesticide use is likely to have beneficial effects on people's health, livestock production, water and air quality, biodiversity, and wildlife welfare at a regional level. They argue that returns on investments in IPM-FFS cannot be appraised until the pesticide-use externalities have been taken into account and properly quantified. (For a global review of the FFS experience, see Braun *et al.* 2006.)

In general, it is recognised that mono-disciplinary studies with pre-determined objectives are no longer considered sufficient to evaluate development interventions centred on people's empowerment. The multi-dimensionality of FFSs demands new methodological and conceptual efforts. First, it has been shown that contributions from several disciplines are needed to address the overall values of programmes that aim at people's empowerment. Second, both the function and agency of evaluation have been highlighted. While some authors argue that external evaluation contradicts the FFSs' core aim to transfer power to users (for example, Bartlett in this issue), others have been more concerned with the inability of conventional methods to capture unpredictable but relevant effects, thereby reducing the relevance of the findings for improving programmes (see, for example, Murray 2000). Participatory methods, however, are regarded positively as remarkably flexible: that is, able to be adapted during their application to the specific context and to increase accountability to users (Murray 2000). User participation in evaluation processes can have the functional role of increasing the efficiency of the programme being evaluated by providing useful feedback. In addition, their participation may be intended to empower users by prioritising the learning inherent in the evaluation process; this application has been defined as empowering participation (Fetterman *et al.* 1996).

However, there are also authors who question the methods and practices of participatory evaluation for their generalised lack of rigour and objectivity. They argue that the assumption that participation can lead to people's empowerment is naive. Cooke (2002) explains how the acts and processes of participation can reinforce injustice and any imbalance in existing power relations, if the complex manifestation and dynamics of power are not understood. He claims that the currently available examples of applied participatory approaches can only reveal but never challenge power inequalities, and that therefore a 'misunderstanding of power underpins the participation discourse'.

The debate about how an evaluation should be assessed, what exactly should be assessed, and who should carry out the assessment is far from being solved, and probably there will always be divergent opinions; but it is clear that a broad array of evaluation approaches and methodological innovation are required to follow the evolving concept of evaluating development. In this article, we report on the case of an evaluation of the ecological, agronomic, social, and human outcomes associated with IPM-FFSs in India, using a range of methodologies from conventional to participatory. The aspects of sustainability addressed by the research were: (1) the occupational health of male and female farm workers engaged in the handling and application of pesticides; (2) the changes in farmers' agronomic practices, particularly in input use, determined by the adoption of IPM; (3) the reallocation of labour associated with the introduction of IPM tasks, and its gender implications; (4) the ecological impact of the emissions released into the environment by conventional, integrated, and organic cotton cultivation; and (5) the overall effects of the IPM-FFSs on livelihoods. Secondary aims of the research were to assess the role of participatory approaches to evaluation in enhancing the relevance of the evaluation, and to test innovative applications of the methods used.

The research framework, including methods for data collection and analysis, is briefly presented. The findings, published elsewhere, are summarised. We proceed to define the degree to which each method was participatory and at what stages farmers were involved. Finally, a critical review of the limitations and strengths of the methods used is reported.

Study design

Research methods

Data were collected between 2002 and 2004 from 20 villages in the cotton-growing states of Andhra Pradesh, Maharashtra, and Karnataka, to assess the short-term effects of IPM-FFSs up to two years after the intervention. The overall design followed the Double Difference (DD) model, identifying the FFSs as the treatment variable and studying the pre- and post-treatment situation for farmers who attended IPM-FFSs and farmers who did not (control). The control farmers were living either in the IPM-FFS villages (first control) or, in some cases, in villages from the same agro-ecological zone but where IPM-FFSs had never been conducted (second control). The matching criteria used to select control farmers were landholding and cropping patterns. The importance of using the DD model in assessing the impact of FFSs rests on the non-random (that is, purposive) selection of FFS participants. In many cases, FFS programmes used explicit criteria to select participant farmers, such as cropping pattern or high pesticide use. Involuntary bias related to villagers' different socio-economic status can lead, and in some reported cases has led, to a higher participation by the more progressive, educated, and wealthy farmers of a village. In the case treated here, it would be incorrect to attribute observed differences between FFS farmers and control farmers entirely to the FFS intervention, since they could be linked also to the characteristics of the selected groups. In this case, for instance, the involvement of women was actively promoted, with the result that, in 2004, in the villages where gender-sensitisation modules were implemented, the participation of women increased to 50 per cent, and across the project areas reached an average of 20 per cent. In order to generate gender-sensitive findings, the

Table 1: Analytical methods used in the evaluation

Method	Analytical approach
Health monitoring	Linear trend analysis (frequencies analysis and chi-square test)
	Multivariate analysis (multiple linear regression)
Farming System	Two-way ANOVA with time as repeated measure
Analysis and Labour	Canonical correspondence analysis
Allocation analysis	
Life Cycle Analysis	Canonical discriminant analysis
Sustainable	Consensus analysis
Livelihood Approach	Wilcoxon Matched-Pairs Signed-Ranks Test
	Step-wise and canonical discriminant analyses
Photo visioning	Text analysis

evaluation study was conducted in selected villages where women's and men's participation were nearly equivalent. In order to control for the bias introduced by this non-randomised sampling of the overall village population, differences between the treatment and the control group were analysed in terms of changes between two points in time and not in terms of absolute differences. Baseline data sets (pre-situation) were collected before the start of the FFSs, whenever needed and possible. The use of the DD model as the overall design also proved useful in revealing spill-over effects from participants to neighbouring farmers.

The research used the following range of methods.

1. A *health self-monitoring tool* within 24 hours of a spray event, based on a list of 18 signs and symptoms of acute pesticide poisoning and details of the field operations executed. The monitoring was repeated before and after the IPM-FFSs. The control case was initially included in the study; however, the response from the farmers not associated with IPM-FFSs was poor, and the data collection for the group was soon suspended (Mancini *et al.* 2005).
2. Questionnaire-based *Farming System Analysis*, including labour use. The labour questionnaire was developed using the FAO Socio Economic and Gender Analysis (SEAGA) tool (Mancini *et al.* in press).
3. *Life Cycle Analysis* (LCA) (Guinee 2002) was conducted in 15 conventional farms and ten IPM cotton-growing farms. An additional 12 certified organic farms were included in the study to inform upcoming policy on organic farming. Data on input use at household level were retrieved from farm records and in interviews with farmers. This component included only post-IPM-FFS information.
4. *Sustainable Livelihood Analysis* using the five capitals concept: financial, human, social, natural, and physical. This component included the FFS and control cases; pre information data were collected on a recall basis (Mancini *et al.* 2007).
5. Photo-based self-reporting of participants' perceptions (also known as *photo visioning*, see www.photovoice.com) of outcomes and impact. The results of the photo visioning have not been published, but its methodological contribution is discussed.

Data analysis

The five methodologies generated both quantitative and qualitative data, which were analysed using parametric and non-parametric statistical methods (Table 1).

Main findings

The evaluation generated findings on the outcomes (change in practices, labour allocation, and yield) and impact (environmental pollution, human health, livelihoods) of cotton-system IPM-FFSs.

Health self-monitoring

The health self-monitoring study documented that a strikingly large majority (84 per cent) of the monitored spray events led to mild to severe pesticide poisoning. The mainstream literature focuses on the people applying the pesticides; our findings show that the handling of pesticides in order to prepare chemical mixtures and refill tanks can also be extremely risky. In the studied communities, as in general in south Indian villages, these operations were mainly performed by women, who experienced a level of poisoning comparable to that of the men. Marginal farmers belonging to low-income classes suffered ten-fold more poisoning than larger land owners, perhaps because of their lower nutrition status and weakness induced by other diseases. Acute pesticide poisoning therefore affects a much larger population than the actual spray operators.

Farming System Analysis

The analysis of the agronomic practices in conventional and IPM-converted cotton systems showed that the current use of pesticides in cotton is largely superfluous; training in alternative plant-protection measures (IPM-FFSs) resulted in a drastic reduction in pesticide use (78 per cent) without compromising crop yields. The adoption of IPM had no consequences for the overall labour requirement of cotton cultivation, nor the total time spent on plant protection, but the types of task performed to control pest damage changed. A time shift from pesticide application to IPM tasks was reported, resulting in a higher contribution of female work to plant protection. The availability of female family workers was shown to be a factor that might limit IPM adoption.

Life Cycle Analysis

Conventional cotton cultivation pollutes the environment, causing loss of biodiversity, water contamination, increased carbon emissions, acidification, and eutrophication. The practices responsible for these adverse effects are the use of synthetic inputs and carbon-depleting operations such as the burning of organic matter in the field. Organic management was shown to reduce these negative effects to negligible levels, while IPM reduced the negative impacts associated with the use of pesticides, but not the global-warming potential. The close-to-zero environmental impact of organic farms in the case studied was achieved at the cost of substantially lower productivity than conventional cotton farms (20 per cent). IPM farms achieved yields higher than conventional farms (30 per cent).

Sustainable Livelihood Analysis

The gains reported above – better health and environment, reduced cost of cultivation, escape from debts, and increased ecological knowledge – were also those perceived by farmers as enhancing their livelihood. However, farmers also expressed a clear appreciation of the increased social capital, in terms of stronger networks, norms, and social trust that facilitate co-operation and co-ordination for mutual benefit, which they associated with attending

IPM-FFSs, which was perceived as a resource to achieve better development and control over their own destinies. Specifically, the social benefits were described in terms of increased collaboration between villagers and stronger connections with agricultural officers and village authorities. Collective action in the Indian rural society studied takes place typically in informal networks, rather than in the official groups registered to receive the financial incentives given by the government. It has been shown elsewhere that communities with higher collective action are more inclined to act for mutual benefits (Krishna 2003). The evaluation study also revealed that the process of personal growth stimulated by participation in the IPM-FFSs was particularly relevant to women, and it confirms the importance of increasing women's access to educational programmes.

Photo visioning

The pictures taken were grouped by the farmers in photo-visioning workshops, at which the photographs were analysed into clusters, connected by causal or explanatory linkages. This process generated new insights into the occurrence of and the relations between unpredicted impacts, such as improvements in animal health and production, as well as in non-target crop management.

Methodological appraisal of the evaluation

Some study components aimed to provide in-depth information concerning a selected aspect of the evaluation and therefore used methods selected to *describe* the behaviour of a small group of farmers/farms – the health monitoring, and the photo visioning. Findings obtained through these descriptive methods can be *extrapolated* with caution to a larger population who might be considered similar (but for whom there is no direct evidence). The Farming System Analysis, the LCA, and the SLA on the other hand aimed to *explain* the behaviour of a representative group of the total population and to allow for *generalisation* of the findings. Given the different informational objectives of the methods of enquiry used, we now turn to assessing the degree of participation, data validity and accuracy, and other strengths and weaknesses of each method. The participatory methods are also evaluated in terms of their educational value, empowerment, and contribution to gender analysis.

Defining the degree of participation in the methods used

The extent to which each method used in this evaluation was participatory is defined here according to the four criteria described by Lawrenz and Huffman (2003): (1) type of evaluation information collected, such as defining questions and instruments; (2) participation in the evaluation process; (3) decisions concerning the data to provide; and (4) use of evaluation information. 'Outsider' evaluators defined the study objectives and selected the research tools in each case. Farmers' involvement in the decision making pertaining to the four steps of the methodology is outlined in Table 2. The LCA and Farming System Analysis represent the non-participative end of the spectrum, being based on questionnaires developed from the reference literature and preliminary diagnostic studies in the field, and leading to findings that were difficult to share directly with the local communities. The health-monitoring study had a primary research focus; however, it also had the additional educational value of being built into the FFS curriculum and therefore was coupled with access to a viable alternative to the use of pesticides. In the SLA, farmers defined the meaning of the concepts that framed the enquiry, and decided on the type of information to be collected within the given framework. The closest to full participation, or

Table 2: Participation of farmers in decision making and programme evaluation

Method	Time of investigation in relation to FFS	Decision on information to be collected	Farmers' participation in evaluation implementation	Format of data collection	Users of evaluation information*
Health monitoring	During and Post (1 year)	Outsiders	Required	Pre-developed visual format	Programme staff and donors, researchers, national policy makers, farming communities
Farming System Analysis	Pre and Post (1 year)	Outsiders, farmers (partial involvement)	Required in the planning phase	Questionnaire Focus-group discussion	Researchers (programme staff and donors, national policy makers)
Life Cycle Analysis	2nd year after	Outsiders	Not required	Questionnaire	Researchers (policy makers)
Sustainable Livelihood Analysis	1st and 2nd year after	Outsiders, farmers	Required	Open-question interviews	Farming communities (programme staff and donors, researchers, national policy makers)
Photo visioning	1st year after	Farmers	Required	Photos	Farming communities (programme staff and donors)

*In the column, the actual users of the findings are indicated, and in brackets potential users are given.

empowerment evaluation, was the photo-visioning study, which was entirely carried out by farmers. The information collected was processed during farmers' workshops and translated into a collective commitment to move towards more sustainable farm management.

Strengths and limitations of the conventional methods used (LCA and Farming System Analysis)

The LCA generated comparable and repeatable results based on regional- and global-scale indicators. An objectivist approach was required in order to ensure an impartial evaluation and allow for generalisation. It generated insights relevant to policy advisers and cotton producers regarding strategic choices for minimising pollution at national level. It is possible to do scenario analyses of LCA applications in order to predict long-term trends, and very often environmental policies have been shaped by the findings of an LCA. However, modelling simplifies realities in theoretical systems that can be quite far from the empirical situation and do not take into consideration the effects of local circumstances. Also, data collection can be rather difficult and expensive if extended, for example, to water quality.

The questionnaire used for the Farming System Analysis was developed in consultation with the farmers and therefore sufficiently comprehensive and representative of the specific agronomic context. The limitations experienced during the survey were in relation to memory recall and the degree to which respondents were willing to give accurate information (bias introduced by the respondents). In conventional methods, respondents rarely have a direct stake in the results and, while this might ensure higher objectivity, it can also result in lower motivation to provide accurate and precise information. For instance, in the context of agrarian subsidies and official assistance programmes, farmers might have reasons to over- or under-report information on yields, income, input use, or production costs. The risk of biases introduced by the respondents is thus shared by all methodologies that place people as main informants, regardless of the degree of participation.

Strength and limitations of participatory methods used (health monitoring, SLA, and photo visioning)

Evaluation validity Beneficiary Assessment (BA) – 'an approach to information gathering which assesses the value of an activity as it is perceived by its principal users' (Salmen 1995) – has become a widely recognised way to assess the validity of development interventions. The two examples of BA reported in this study – the SLA and the photo-visioning methods – showed that the livelihood concepts used were meaningful to the farmers, establishing the inner validity of the enquiry method, but it also showed that the overall evaluation framework, developed externally, had overlooked areas of impact relevant to farmers. (It might be important to note here that the researchers had an extended affiliation with FFS programmes and therefore were considered to be well informed about on-going debates on FFS impacts.)

Development practitioners agree on the importance of understanding the complexity of people's livelihoods and the inter-linkages that make impact in one area likely to be felt in others. SLA was instrumental in the analysis of these relational aspects and systemic interactions. For instance, its application in two cotton-growing regions revealed that the implications of the knowledge acquired in the FFSs were determined by the farming context. Specifically, during a season favourable to cropping, IPM-FFS farmers achieved better yields at lower cost of production, and in the circumstances of a drought were better able to minimise financial losses than the control farmers in the non-IPM-FFS villages. Given that drought in the study area is fairly regular, the study outputs suggest that a stronger focus on livelihood vulnerability and resilience would increase IPM-FFS efficiency.

However, the validating function of participatory evaluation also meets firm opponents. Mosse (2001) asserts that in participatory approaches local knowledge is often manipulated and instrumentalised to legitimate project objectives and the decisions already made. He regards people's knowledge as a political artefact, reflecting the expression of existing power relations among the participants, as well as participants' strategic adjustment of their needs to match project deliverables. Even though this seems to indicate areas for essential improvements, rather than invalidating the theory of participation, the evaluation reported in this article noted that dealing with existing unequal power relations to facilitate social inclusiveness and gender equity in IPM-FFSs was a demanding task. In FFS weekly meetings, conflicts of interest among participants often surfaced and remained unaddressed. The participatory evaluation tools and processes used in this study were not able to investigate problems arising from the representativeness of participation in IPM-FFS implementation. The sampling procedure of the evaluation took care to ensure a fair representation of women and poor farmers; however, to what extent their will and opinions were evenly reflected in the findings is difficult to state for the more participatory methods such as the photo visioning.

Evaluation accuracy It has been observed that people who are motivated by feelings of ownership of the process provide more complete and accurate information. For instance, pesticide-poisoning studies based on questionnaires have shown much higher poisoning figures than hospitals registries, and directly observed poisoning rates have been recorded that are even higher than recalled information provided by farmers in questionnaires (Kishi 2002). This suggests that measured health data through participant observation or participant farmers' monitoring could increase data accuracy and contribute to a better understanding of the issue.

In this evaluation, the IPM-FFS farmers were predisposed to disclose reliable information, assured by the trustful relationship that linked them to the facilitators. However, extensive and sympathetic interaction between project participants and evaluators can be seen as a source of possible bias. In the health monitoring, for instance, it was considered that the respondents, extensively trained in poisoning diagnosis, were likely to be involuntarily biased towards reporting higher levels of poisoning. The researchers subsequently estimated the level of over-reporting to have been between 17 per cent and 38 per cent, based on three dummy symptoms introduced deliberately into the reporting format. In the SLA, a generally positive attitude concerning change over time was recorded for the male respondents who had been FFS participants; even though the variation among respondents was high (suggesting reliability), the risk of bias from FFS-affiliated respondents remains a danger.

The language used to conduct participatory methodology is unmistakably the one spoken by the local communities. However, if the stakeholders involved speak different languages and/or the scope of the evaluation goes beyond evaluation at community level, some of the depth of the findings can be lost in the analysis. The senior researchers relied on translated texts, as the investigation was carried out in the local languages, and this might have limited the accuracy of the transcribed information.

Empowering evaluation? The opportunities for empowerment provided by participatory evaluation are entailed in the process through which such evaluations proceed – an experiential learning cycle. Starting from a concrete experience, participants conceptualise and apply new principles that can lead to a mental transformation (Percy 2005). This evaluation study showed that the deeper understanding of the occupational hazard of handling pesticides indeed induced a change in the FFS participants' attitudes towards pesticides.

There are other features of participatory evaluation that can be seen as a means to increase the representation of people's own meaning and purpose. Visual aids, for instance, assist the co-generation of knowledge through non-verbal representation and dialogue. The photo-visioning method in particular helped people less confident in verbal communication or drawing techniques to express themselves, and to convey deeply textured meaning through the stories told around the photos taken.

The health monitoring, however, provided an instance in which participatory principles were traded off against pragmatic aims. In the periodic meetings organised during the self-health assessment to review monitoring forms, various health issues were surfaced but not properly followed up because they went beyond the specific research outline. In a truly empowering participatory evaluation, local control and autonomy would have overridden the need to comply with a set of fixed objectives. Asking the people to participate in the project's agency limited the empowering scope of the process through which the method was applied.

Finally, participatory methodologies can be rather time-consuming (the health monitoring), adding an extra task to the already demanding daily schedule of farmers, particularly in the case of women farmers. Their application at regional level can be difficult. Participatory techniques are usually applied with a small number of people to ensure in-depth interactions, and to study specific local outcomes. However, a participatory study method applied with a research purpose can generate results through a rather standardised procedure and statistically comparative frame, as was the case with the SLA study.

Participation and gender

Participatory techniques are not gender-sensitive *per se*, but they can be if accompanied by an approach sensitive to diversity of ethnicity, gender, age, class, religion, and culture. The FFS model, with its emphasis on peer review and informed decision making at individual, group, and community levels, might leave the priorities of weaker groups unexpressed or unaddressed (Guijt and Kaul Shah 1998). Prior to the evaluation, substantial time was dedicated to focus-group discussions and to the eliciting of labour calendars with women and with men, and to periods of participant observation. This work shaped the definition of data requirements and the data-collection procedures. Data disaggregated by gender and social class were collected for every evaluation domain, except the LCA. The following study dimensions would have been missed without the participation of women in the overall evaluation.

The list of operations surveyed by the labour-allocation analysis included a number of secondary operations performed by women that are frequently overlooked. Their inclusion allowed for a more accurate estimation of the labour requirement in cotton farming. In turn, the study showed that the availability of female labour is a factor influencing the rate of IPM adoption, invalidating the misconception that plant-protection measures are entirely a male domain. The gender-labour analysis, paired with participant observations, also laid the foundations for the health monitoring, by revealing the separation of roles in pesticide application. The SLA revealed that building human capital was particularly relevant for women, who developed, through the IPM-FFSs, the confidence to voice their needs and opinions. It is important to realise that the introduction of new technology or management has social and gender implications that need to be taken into account in evaluation studies.

The evaluation did not analyse the social class, religion, or any other cultural variables of the women attending IPM-FFSs in relation to the total village population, and it cannot therefore offer conclusions on the issue of social inclusiveness in the IPM-FFSs. Capturing these aspects was beyond the initial scope of the evaluation and not within the reach of the methods used. However, the study has provided evidence that IPM-FFSs are relevant to women and poor farmers, and that targeting strategies, including facilitators' training in problem-solving skills and gender-sensitive approaches, are needed to overcome barriers to their participation.

Conclusions

The evaluation study applied two methodological approaches – conventional and participatory – which made different, but equally rigorous, claims to knowledge. The former provided an objective measurement of selected environmental and social impacts generated by different cotton-production systems. The findings generated are particularly relevant to refine the technical content of the IPM-FFS curriculum, as well as national plant production and protection policies to minimise pollution and sustain farmers' returns. The study also revealed that the same intervention had different impacts on the livelihoods of women and men, and of wealthy and poor farmers, and that the value of the experience can also be captured in terms of the farmers' own frames of reference.

An effort has been made in this article to assess the impact of the methods, and to make a transparent and systematic interpretation of findings which have been validated with reference to both the experience of farmers and scientific peer review. However, we consider that when the object of study is the messy real world, no findings can be considered final or absolute.

References

Braun, A., J. Jiggins, N. Röling, H. van den Berg, and P. Snijders (2006) 'A Global Survey and Review of Farmer Field School Experiences', Nairobi: International Livestock Research Institute, and Wageningen: Endelea.

Cooke, B. (2002) 'The case for participation as tyranny', in B. Cooke and U. Kothari (eds.) *Participation: The New Tyranny?* London: Zed Books.

Feder, G., R. Murgai, and J.B. Quizon (2004) 'Sending farmers back to school: the impact of farmer field schools in Indonesia', *Review of Agricultural Economics* 26 (1): 45–62.

Fetterman, D. M., S. J. Kaftarian, and A. Wandersman (1996) *Empowerment Evaluation: Knowledge and Tools for Self-assessment and Accountability*, Thousand Oaks, CA: Sage Publications.

Guijt, I. and M. Kaul Shah (1998) *The Myth of Community: Gender Issues in Participatory Development*, London: Intermediate Technology Publications.

Guinee, J. B. (2002) 'Life Cycle Assessment: An Operational Guide to the ISO Standards', The Netherlands: Ministry of Housing, Spatial Planning and Environment (VROM) and Centre of Environmental Science, Leiden University (CML).

Kenmore, P. (1996) 'Integrated pest management in rice', in G. Persley (ed.) *Biotechnology and Integrated Pest Management*, Wallingford: CAB International.

Khalid, A. (2002) 'Assessing the long-term impact of IPM Farmer Field Schools on farmers' knowledge, attitudes and practices. A case study from Gezira Scheme, Sudan', in *International Learning Workshop on Farmer Field Schools (FFS): Emerging Issues and Challenges*, Yogyakarta, Indonesia: CIP-UPWARD.

Kishi, M. (2002) 'Farmers' perceptions of pesticides, and resultant health problems from exposures', *International Journal of Occupational and Environmental Health* 8 (3): 175–81.

Krishna, A. (2003) 'Understanding, measuring and utilizing social capital. Clarifying concepts and presenting a field application from India', *CAPRi Working Paper* No. 28, Washington, DC: International Food Policy Research Institute.

Lawrenz, F. and D. Huffman (2003) 'How can multi-site evaluations be participatory?', *American Journal of Evaluation* 24 (4): 471–82.

Mancini, F., A. van Bruggen, J. L. S. Jiggins, A. Ambatiputi, and H. Murphy (2005) 'Incidence of acute pesticide poisoning among female and male cotton growers in India', *International Journal of Occupational and Environmental Health* 11 (3): 221–32.

Mancini, F., A. H. C. van Bruggen, and J. L. S. Jiggins (2007). 'Evaluating cotton integrated pest management (IPM) farmer field school outcomes using the sustainable livelihoods approach in India', *Experimental Agriculture* 43: 97–112.

Mancini, F., A. J. Termorshuizen, J. L. S. Jiggins, and A.H.C. van Bruggen (in press) 'Increasing the environmental and social sustainability of cotton farming through farmer education in Andhra Pradesh, India', *Agricultural Systems*, in press.

Mosse, D. (2001) '"People's knowledge", participation and patronage: operations and representations in rural development', in B. Cooke and U. Kothari (eds.) *Participation: the New Tyranny?* London: Zed Books.

Murray, P. (2000) 'Evaluating participatory extension programs: challenges and problems', *Australian Journal of Experimental Agriculture* (40): 519–26.

Percy, R. (2005) 'The contribution of transformative learning theory to the practice of participatory research and extension: theoretical reflections', *Agriculture and Human Values* 22: 127–36.

Pontius, John, Russell Dilts, and Andrew Bartlett (2002) *From Farm Field School to Community IPM. Ten Years of IPM Training in Asia*, Bangkok: FAO Regional Office for Asia and the Pacific.

Rola, A., S. Jamias, and J. Quizon (2002) 'Do farmer field school graduates retain and share what they learn? An investigation in Iloilo, Philippines', *Journal of International Agricultural and Extension Education* 9 (1): 65–76.

Salmen, L. F. (1995) 'Beneficiary assessment: an approach described', in *Social Assessment Series* No. 23, Washington, DC: World Bank Environmental Department.

Shetty, P. K. (2003) 'Ecological Implications of Pesticide Use in Agro-ecosystems in India', Bangalore: Environmental Studies Unit, National Institute of Advanced Studies, Indian Institute of Science Campus.

Engaging with cultural practices in ways that benefit women in northern Nigeria

Annita Tipilda, Arega Alene, and Victor M. Manyong

This study explores the intra-household impact of improved dual-purpose cowpea (IDPC) from a gender perspective, in terms of productivity and food, fodder, and income availability, the impact of which is linked to the income thus placed in the women's hands. Surplus income is important in providing food and nutritional benefits to the home, particularly during periods of risk. More importantly, income generated through the adoption of improved cowpea varieties has entered a largely female domain, where transfers of income reserves were passed on between women of different ages, with significant impact in terms of social and economic development. However, the technology has strengthened the separation of working spheres between men and women. Future technologies should, from the outset, explore provisions existing within the local rubric, to focus on women with the aim of expanding their participation in agriculture with the associated benefits to their families.

Introduction

The intra-household distribution of benefits from increased agricultural productivity largely depends on the relative bargaining power of household members. In the context of asymmetric gender relations, investments in human capital and social welfare are often dependent on the ability of the woman to influence decisions within the household. Increased income in women's hands can have the potential for egalitarian shifts in the distribution of benefits and influence in decision making. In addition to this, studies have shown that in sub-Saharan Africa it is women's control over key economic resources, rather than mere economic ownership or participation, that is critical to their power within the family (Quisumbing and Maluccio 2000; Haddad *et al.* 1994; Blumberg, cited in Kibria 1995; Dey 1992). When women not only earn but also control the use of their income, they can use it more effectively in bargaining, with the implicit threat of withdrawing from the household economy.

This likelihood calls for an intra-household analysis of the gender-differentiated impact of agricultural programmes that aim to improve agricultural productivity and livelihood outcomes.

This article is based on a qualitative study of intra-household impact of improved dual-purpose cowpea (IDPC) technology from a gender perspective. Impacts were assessed using the sustainable-livelihoods framework, namely, impact of IDPC on human, natural, financial, social, and physical capital (Adato and Meinzen-Dick 2002). Given that complex social and cultural concepts of perceived and actual contributions were involved, a multi-disciplinary approach was adopted. In this case, both qualitative and quantitative approaches were used, in order to understand the intra-household impact. Qualitative research methods have become increasingly important modes for enquiry into complex social phenomena. Qualitative research has the flexibility that allows respondents to offer interpretations and choose themes which broaden the scope of the research and deepen understanding of the social processes. It helps researchers to understand the culture, perception, attitudes, and opinions of people, and to explore their interpretations of different phenomena, that is, to gain an insider's perspective.

This study sought to assess perspectives in relation to four questions:

- Has income increased from implementation of IDPC technology and, if so, who controls the additional income? And who benefited from it ultimately?
- Has food security and nutrition improved as a result of the adoption of IDPC?
- How are other benefits from implementation of the new technology distributed within the household, the community, and between women and men?
- What is the impact on gender relations?

Finally, the study provides recommendations for future International Institute of Tropical Agriculture (IITA) research in benchmark areas in northern Nigeria.

Asymmetry and intra-household dynamics among the Muslim Hausa

Among the Muslim Hausa of northern Nigeria, socio-economic organisation is characterised by a system of stratification, based on occupation, wealth, birth, and patron–client ties. Wealth gives its possessor a certain amount of prestige and power, especially in forming ties of patronage. Status is determined by one's family's ability to contribute to social and religious events, and this in turn influences the family's ability to participate in wider decision-making processes. Gender is strongly moulded by the existing patriarchal system, which determines the power relations within the households and the bargaining power of household members through various mechanisms. Key among these are family, marriage, inheritance patterns, wealth, and occupation

The fundamental unit of Hausa society is the household compound (*gida*), headed by a senior male owner of the compound (*maigida*), and often by his adult sons, who are not yet wealthy enough to form their own compounds and who often engage in farming with the household head in a unique family farming system (*gandu*). Farming activities are organised among related men within a household compound. In this system, the senior member (household head) organises production on one or more farms that are used to grow food or feed for the household during the year. This often involves family labour, which is usually paid for in cash or in kind, often in the form of grain such as cowpea. Cowpea is grown by virtually all farmers, either for consumption or for sale. The actual ratio of cowpea grown for any individual often depends on unforeseeable circumstances and productivity of the farm.

Occupational specialities, such as sewing leather wallets, are pursued on a more individual basis. The Muslim Hausa wife is obliged to provide labour for the preparation of food, child care, and general domestic chores. Because women are secluded within the household, a wife is not expected to work in the fields or to fetch water. If a husband cannot provide food from his own farm, he must purchase it from the open market for his family – often with cash borrowed from his wife (from her petty trading with women in other households). Female seclusion is, therefore, a major determinant in the gendered division of labour, resulting in economic imbalance between the two sexes. Men maintain the household for the benefit of the family unit, while women are expected to acknowledge the shelter provided by adopting appropriately modest behaviour. Households vary in the extent to which women are secluded. Among the elite, full seclusion is the general rule. Partial seclusion or no seclusion also occurs.

Within their separate worlds, men and women are free to pursue their own economic activities, as long as the above-mentioned obligations are met. Men engage in contract labour; the farming of cash crops such as cotton or groundnuts; trading in cloth, kola nuts, and other general items of trade; and transportation. Or they may start their own businesses within the wider economy. Women's economic activities consist mainly of trade between compounds, resulting in what has been called the honeycomb trade in Hausa land, with many busy little cells of sheltered women, involved in a wide but relatively hidden network of trading relationships with one another (Hill 1969; Schildkrout 1986; Renne 2004). Women also process crops within the compound, and much of this work is remunerated.

Because women are expected to be wives and mothers and to remain secluded within the compounds of their husbands, little value is placed on girls' formal education. Although females' understanding of the Koran and the basic tenets of Islam is deemed important, girls mainly attend a Koranic school under the tutelage of a local scholar, and only for a short period of time. Separation and seclusion from the outside world are most important while a woman has reproductive potential. Respectability requires that this reproductive potential be under appropriate male control. After menopause, women are allowed much greater freedom; with some becoming much more active in the wider economy, occasionally amassing significant wealth (Coles 1990; Schildkrout 1986).

Within the household, a husband has no control or right to the woman's income and neither can he oblige her to contribute to its upkeep. It is the husband's duty to provide all the grain and meat (as well as firewood) required by his dependants, and to give his wife a daily sum in cash (*Kudin cefane*) for the purchase of other ingredients, which are usually bought from other secluded wives. If a husband cannot provide for his dependants from his own granaries, then it is his duty to buy the necessary grain himself or to provide cash. In many instances, women have reported lending to their husbands at interest, or selling grain to their husbands at planting season.

A Hausa compound is typically constructed with high mud-brick walls with entrance only through a small entry hut (*zaure*) which opens on to the street and in which the compound head greets visitors. The composition and number of individuals living in a compound, as well as their relationships with one another, are often fluid and changing.

Improved dual-purpose cowpea-breeding programme

The leguminous cowpea (*Vigna unguiculata* [L.] Walp.), a popular crop in West Africa, brings multiple benefits to rural communities. Its grains are a major food staple for rural and urban households, and a source of cash when sold raw or processed into various products. Women carry out the processing and, as such, cowpea plays a critical role in their empowerment. Cowpea fodder is a source of protein for animal feed and is easily stored and sold during the dry season for additional cash. Healthy livestock in turn produce more manure, which helps to sustain crop production. In emergencies, livestock serve as a quick source of cash. Cowpea fixes atmospheric nitrogen and contributes to soil fertility for the benefit of the subsequent cereal crop in the rotation. It also acts as a trap crop which stimulates the suicidal germination of *Striga harmonthica*, a parasitic weed that can cause the entire loss of cereal yields.

Despite its popularity and multiple roles in the farming systems, productivity levels in farmers' fields remain low. Some of the constraints are high pressure from pests and diseases, poor management practices, lack of inputs, and inefficient extension services. In order to address some of these constraints, the International Institute of Tropical Agriculture (IITA), in collaboration with the International Livestock Research Institute (ILRI) and partners in the national agricultural research system, developed a basket of technological options that included high-yielding, improved dual-purpose (grain and fodder) varieties, new crop spatial arrangements that alternate rows of cereals with those of cowpea, and a minimal use of external inputs. Dissemination of improved cowpea varieties started in 1993/1994, beyond the initial experimenting villages, with the assistance of the state extension services in Kano and Kaduna states of northern Nigeria. In 2004, an adoption and impact study was undertaken to assess and explore the intra-household impact of improved cowpea varieties.

Study area

The study was conducted in Kano and Kaduna states in northern Nigeria, each representing one of two agro-ecological zones: the northern Guinea savannah and the Sudan savannah. Mean annual rainfall ranges from 500 mm in the northern fringes to 1600 mm along the southern boundary. Rainfall is unimodal and allows 75–180 days growing period across the north–south gradient. Major cereals grown are sorghum and millet in the dry semi-arid areas, and maize in the Guinea savannah.

Sample design and data collection

In order to establish a sampling frame for the selection of representative farmers, an initial exploratory field survey was conducted to identify all villages growing IDPC in Kano and Kaduna states. From the 101 villages in Kano and 27 villages in Kaduna that were actually growing IDPC, the list of villages was then stratified against resource-use domains of population and market access. Population pressure and market access are the two most important 'drivers of change' for agricultural intensification and adoption of improved technologies in West Africa (Manyong *et al.* 1996; Kristjanson *et al.* 2005). The varieties of cowpea grown and their role in the farming systems and livelihoods depend mainly on three socio-economic factors: human population density, livestock population density, and access to a wholesale market (for obtaining farm inputs and for sale of produce). Against these stratification variables, with the associated four socio-economic domains,[1] a total of 24 villages growing IDCP were randomly selected from the two states. Given that Kano is an important cowpea-growing state, 16 of the 24 villages were selected from that state.

With the assistance of extension staff, enumerators, and farmers in the village, a list of farmers living in the respective villages was drawn up. Regardless of their status as adopters or non-adopters, 20 farmers were randomly selected from each of the 24 villages. Thus, the survey data were collected from a total of 480 household heads, their wives, and children. The wives of the sample household heads were interviewed, and anthropometric measurements were taken of children aged 12–84 months. Household-level surveys were used to collect data on socio-economic characteristics, crop production and cropping systems, varietal adoption and diffusion process, and benefits and constraints of IDPC. As one of the objectives of the study was to assess the impact of IDPC on intra-household relations, data were collected from both the household heads and their spouses.

Results

Adoption

About 79 per cent of the sample farmers in Kano and 61 per cent in Kaduna had adopted improved cowpea varieties. Improved varieties occupied about 20 per cent of the total land cultivated by the sample households in Kano and about 14 per cent of the land cultivated by the sample households in Kaduna. Considerably higher yields (1166 kg/ha) were reported among adopters in Kano than among non-adopters (498 kg/ha); this represents a 134 per cent increase in yields brought about through adoption of improved cowpea varieties. In Kaduna, adopters also obtained higher yields (1129 kg/ha) than non-adopters (652 kg/ha) – a 73 per cent increase in yields associated with adoption of improved varieties. An increase in yield means an increase in income, food, and fodder.

Livelihood capital assets

Household heads' assessment of changes in their households, productive assets, and human-capital investments following the introduction of improved cowpea revealed that farmers who had been cultivating it had acquired a range of assets and had been able to make investments in education and health care. The incomes obtained from the sale of improved cowpea enabled about 24 per cent of the adopting households in Kano to make human-capital investments in the form of health care, and about 16 per cent were enabled to invest in the education of their children. Furthermore, about 33 per cent of adopters acquired goats, 30 per cent bought sheep, 17 per cent bought bicycles and radios, and about 36 per cent bought clothes. In Kaduna, 20 per cent of adopters constructed metal-roofed houses, 19 per cent bought bicycles, and 55 per cent bought clothes. Adopters in Kaduna also acquired productive assets out of IDCP incomes: about 20 per cent bought goats, 19 per cent bought cows, and 16 per cent bought land.

IDPC technology and household impacts

There has been growing renewed interest in intra-household allocation of welfare in recent years, with the realisation that slight differences in the allocation of scarce resources can have dramatic consequences, particularly on children's and female nutrition, morbidity, and mortality (Haddad *et al.* 1994). Recent studies have rejected the hypothesis that households are unitary, and indeed have demonstrated that consumption among household members varies systematically with their relative contributions and control over such resources – that is, their ability to bargain with the implicit threat of withdrawing these resources. Bargaining power has four determinants: control over resources; influences that can be used; interpersonal networks; and value systems found within the community. Factors that can influence the bargaining process include education, skills and knowledge, the capacity to acquire information, personal networks, and membership of organisations or groups. Several studies have found the bargaining model useful in understanding impacts (Haddad *et al.* 1994; Naved 2000; Quisumbing and Maluccio 2000).

This study confirms previous findings. The Muslim Hausa household is not a unit that links families in subsistence agriculture, in which a household head can call on the labour of junior men and women within the household to work on the farm with no direct remuneration. Households among the Muslim Hausa are non-corporate in most functions, meaning that the household is not a production or a consumption unit and has no conjugal fund linking husband and wife as one financial unit. Relations of production are largely extra-household, with women being mobile between homes through divorce and re-marriage. Thus, measures of impact of improved agricultural production on men, women, and children are linked to the separation of their economic spheres and spaces.

A gender-differentiated study of the household has been particularly useful, given that the impact of increased IDPC and increased resultant income on women's contribution to the creation of market-value products still remains socially invisible, due to their absence at the point at which the value of such products is realised. Disaggregating impact by gender has provided useful data on how IDPC – in contributing income through higher productivity – has altered the lives of women in their household setting, and revealed how women have responded creatively to these external pressures in fashioning satisfactory lives for themselves and their daughters. In turn, these impacts influence the course that younger women will follow later, thus affecting the course of economic development and the path of social change.

Impact of increased income from IDPC

In this section, we discuss the avenues through which income has increased among the wives of adopters as a result of IDPC adoption. The techniques used for data collection included key-informant interviews, focus-group discussions, case studies, trend analyses, impact flow, and historical profile. Triangulation among various qualitative and, where necessary, quantitative techniques was used to validate the information gathered. Although this article attempts to present the perspective of the women, the conclusions are based on a synthesis of these data with the understanding and analysis of the researchers. Scholars of intra-household impact who have applied a gender perspective have used similar techniques to elicit data on the impact of increased use of agricultural technologies on women (Naved 2000). The exact level of income generated by the women in processing the IDPC, or as provided for by the household head (*Kudine cefane*), was found not to be useful, because it was doubtful whether reliable statistics would have been obtained because of the extreme secrecy that surrounds various transactions (for example, in the petty trading and the re-sale of grain for 'own production' to the husband). We found qualitative approaches useful for a society where husbands and wives are often kept ignorant of their spouse's economic affairs, and where there is extreme secrecy concerning exact levels of expenditures.

IDPC did increase income

There was consensus among the spouses of adopters that income had increased as a result of the increased productivity of improved cowpea varieties. Additional income was made available to them by the household head (*Kudin cefane*) from the sale of grain. This income was used to expand or initiate petty trading activities – for example, the sale of Islamic caps, salt, and leather products. Additional grain, available to them within the household, was used in the production of bean cakes for sale, a common activity among Muslim Hausa women. Some of the grain within the household was saved over from the previous season's harvest by the women, and then re-sold to the household head during planting season – a common practice between husbands and wives.

> *I have noticed that after the last two harvests there was additional grain within the household, and the amount of cash he gives me has increased. For me, I had additional income with which I was able to invest in my petty trading. I used additional grain available to the household to process into bean cakes (Kosai), and the rest I re-sold to my husband during the next planting season, for a price. Many women in this area whose husbands have adopted cowpea have been able to do the same. I was also able to purchase a small*

wheelbarrow, which my children can use to transport goods for neighbours to the nearest tarred road for a small price. (Magajiya Fada, Kayawa village, Eiwa Local Government Authority, Kaduna state)

Uses of increased income by the women

Food security and nutrition For the Hausa, the food situation within the household has two distinct periods and is a reflection of the difficulties in maintaining consistent food-consumption patterns:

- *Matsala* refers to the food situation in a household where produce is inadequate and lasts only three to five months after harvest. Adjustments in the daily quantity and quality of meals start shortly after harvest, around January. From January onwards, two meals are generally consumed daily, reduced to one around May.
- *Tararrabí* refers to the situation in households where produce lasts eight to nine months after harvest, and the families concerned start adjusting the number and quality of meals around April/May. Shortly after harvest, breakfast comprises left-over *tuwo* (maize meal) from the previous night, supplemented by porridge and cowpea bean cake. Lunch comprises dishes made from a cereal and cowpea mixture. Dinner, the main meal for most families, consists of *tuwo* with a local soup made from dried and ground okra (*kubewa*), baobab leaves (*kuka*), or rosselle (*yakua*). Fish or occasionally meat may be added. Milk is often consumed at lunch as an accompaniment to millet and sour milk drink (*fura da nono*).

During difficult periods, young women intensify petty-trading activities. Income accrued by women during periods of food scarcity, including income from livestock sales, goes towards supplementing family food stocks. Where several generations co-exist in a household, the older women often assist the men to purchase additional cereals, while younger women enhance family food stocks by purchasing soup ingredients from their petty-trading activities. Income generated by men alone is inadequate to maintain the frequency and quality of meals, and women contribute additional income.

IDPC was reported as having a positive impact. Increased productivity meant increased grain reserves for sale as processed bean cakes and for consumption; with anything remaining being re-sold before the planting season. Unearned income from the household head (*Kudin cefane*) from the sale of IDPC was used to purchase additional foods, or held in reserves such as livestock and enamel plates (*kwano*), which would be disposed of during food-insecure periods. For adopter households considered poorer in the community, *Kudin cefane* (from the sale of grain by the household head) also assisted the women to start up petty trading or expand existing activities, which were important during difficult periods. Such activities generated more income, which enabled the household women to purchase foods on the market.

> *I can only report for our household. I am able to purchase different foods during difficult periods. The women in this household are still able to afford vegetables, especially tomatoes, for our food preparation. We can afford different types of meat. This is special, because we normally could not afford such food. That is the immediate change I have noticed. It is small but it is as a direct result of improved cowpea varieties.* (Safiya Yusuf, Rirnaye Raki village, Bichi Local Governrnent Authority, Kano)

> *previously, we had to eat pap (porridge) during the early planting season, when we had no cowpea grain, and we had to spend cash to purchase seeds on the market for our crops. Now we can afford rice and occasionally some meat and milk. Now during the dry period (crisis period), we may go without fish or meat, but not without some rice or*

vegetables, I am even able to add extra cooking oil in the food. (Ade Idanan, Santa Rago, Bichi Local Authority, Kano)

My household has not been growing improved cowpea varieties, because we have had no access to fertiliser and my husband is poor; but we have noted some households growing improved cowpea varieties in 2003 season. What I can say is that I have heard that the women in the areas growing improved cowpea varieties are able to change their diet and also have grain and food in stock throughout the year. They are also able to meet the household needs more easily. (Wife of non-adopter, Zulai Sale, Malikanch Ikara Local Government Authority, Kaduna)

During the focus-group discussions, indicators which according to the women represented good health, nutrition, and welfare were listed. The IDPC technology was assessed against this list. Among these indicators selected to represent changes brought by IDPC in adopter households were healthy skin without rashes; physically active plump children; children able to attend both Islamic and formal schools; children able to communicate and understand. These perceptions and income invested in food security and the nutritional values attributed to cowpea were tested against anthropometric data. The results were calculated under three groupings of households: two years of adoption of improved cowpea (ADOP2), three years of adoption (ADOP3), and four years of adoption (ADOP4).

Overall, the average anthropometric results indicated scores within ranges of good nutritional status. No significant differences were observed between children of adopters and those of non-adopters for all the anthropometric scores for ADOP2. One significant difference in the weight-to-height ratio was observed for ADOP3, and two significant differences were observed in the weight-to-height and weight-to-age ratios for ADOP4.

These results imply that children's nutritional status was good. This is consistent with perceptions of wives of adopters. These results may show that, as sustainable adoption occurs over time, differences are being observed between children of adopters and those of non-adopters.

Use of income for education Among the Muslim Hausa, the relative importance of formal education is low, particularly for girls. Young children often assist their mothers in petty trading between households and receive only Islamic tutoring. For the few who do receive formal education, it is the boys who are granted the first priority; while the girls are expected to assist the mother in household chores and, at the onset of puberty, be married off.

Some wives of adopters still reported that they were able to contribute to formal educational costs of their children, often the boys.

We have managed so far to have additional income, which has boosted our ability to take our children to both formal and Islamic schools. (Ade Idanana, Santa Rago, Bichi Local Authority, Kano)

Use of increased income for health-care needs For Muslim women in northern Nigeria, access to modern health facilities is largely inhibited by the prevailing social norms of wife seclusion. This has in turn an effect on the health care of young children. A clear indicator of this is the high maternal mortality ratio (more than 1000 maternal deaths per 100,000 live births). Marriage at an early age often compounds the situation, and many women and children are directly affected by harmful traditional medical beliefs and practices. The convergence of all of these factors has resulted in one of the worst records of health for women and younger children (Renne 1996; Renne 2004).

Increased income from IDPC contributed to the meeting of health-care needs of both young women and children. And this was achieved by offering them a choice of modern or traditional care, with the modern care being sought only in extreme cases, where the traditional failed. In focus-group discussions, it was stated that this was particularly so in the case of childbirth difficulties. However, where young children were concerned, increased income meant that they could afford modern health care provided at a nearby dispensary.

> *Women in childbirth traditionally are required to remain at home and receive help within the confines of the home. Where there are complications, money is collected within the household and village, and the woman is rushed to the nearby hospital – often when it is too late. Now that women have increased income, they can plan the timing when improved health care can be accessed when in childbirth. But for children, increased income definitely has helped us to afford immunisation, as well as improved health care.* (Baita village, Wudil Local Authority, Kano)

Impact of increased income on inter-generational transfers Inter-generational transfer of resources between women and their daughters is a common practice among the Hausa Muslims. Women's ability to do this depends heavily on the success of their petty-trading activities. Surplus income from petty trading as a result of IDPC was held in reserves such as livestock (mostly goats), and enamel dishes (*kwano*), which are then handed down to their daughters in preparation for their future married lives. Such transfers of wealth to the daughter have the wider impact of assisting young women to start up their own independent economic activities, as well as increasing their ability to influence decisions. In the event of divorce, which is a common occurrence in difficult marriages, cash held in such reserves helps them to start again, either in a new marriage or back in their parents' home. By contributing to the income-generating activities of the women, IDPC has reinforced the inter-generational transfers between women.

> *Most of the income that women earn is used within the household. But I must tell you of a very important practice. We, particularly as older women, invest the money in* kwano, *when our girls are starting to mature. When my daughter gets married, I will hand over some assets to her, and she can survive. Young girls are not allowed to trade or be involved in farming or petty trading. They must wait until they are older before they can be allowed to contribute actively within the household. In the event that she is divorced, or is unable to have children, she can still have a reserve, until she joins a household with co-wives and their children. It is very important for us women to help our children. And in my opinion (and I think the other older women agree), any increased income within the woman's hands, particularly older women, is a contribution for the next generation. And IDPC reinforced this.* (Aisha Garba, Senta Rago, Bichi Local Government Authority, Kano)

By providing a basis on which women can start up or increase their income-generating potential, IDPC has provided income with which they are also able to purchase gifts for weddings, naming ceremonies, and for other women. This is seen as cementing bonds between women (*Adashi*) and, as such, ensuring a safety net during times of hardship, when such favours would be reciprocated.

> *In terms of the participation of women, there has not been much change at the household level, but at the community level there are better relations for the women, because extended families are assisted, and this means more peace and respect. When your in-laws and fellow neighbours are helped, you have built yourself a bridge over a difficult period in the future. And in this process women are able to participate in the decision*

making in which neighbour and family issues are concerned. That is the only notable change I have seen in our few households growing improved cowpea varieties. (Hauwa Liman, Rimaye Rak, Bichi Local Government Authority, Kano)

More income, more savings, more collective action

In villages around Kano, women have been able to start up micro-credit groups. For some women, membership of existing groups is gained through the ability to afford monthly contributions. For others, women have come together to organise their income into joint savings groups. Some credit groups are run between secluded women, based on trust, and linked through their children. For others, the women are allowed by their husbands (under appropriate conditions) to meet with other women, and discuss activities and contributions. Such credit groups or 'merry-go-round' groups (as they are commonly referred to) are important in starting up or intensifying income-generating projects.

From what I can see, these women of adopting households in our village here in Kano are now able to have a little money, which they contribute to a common pool among the women who form groups in this ward. This money is then used either for community activities, or is given to women who contribute. Some also have their own little groups in which they sew woven Islamic caps together. There is also one group where they were able to buy a sewing machine, and they together are working on selling Islamic wear such as kaftans. (One wife of a non-adopter had this to say on her observation of spouses of adopters.)

The wives of adopters have access to large social networks through the petty trading business and self-help women's groups that they gained membership to with the additional income from IDPC. (Su Waiba Nadabo, Malikanchi Ikara Local Government Authority, Kaduna)

Effect of increased income on women's status and gender relations

An understanding of gender relations was obtained by questioning both women and men about gendered ideologies and behavioural roles, and the changes that have occurred with IDPC, leading to an estimation of their relative power – defined as access to, benefit from, and control over significant resources. The seclusion of women and the strict separation of male and female spheres did not necessarily imply bad gender relations or the subordinate status of women. Previous studies have found that, where distinctive ideologies of men and women occur, a balance of power is most likely to be present, and gender relations are often good. Beyond the increase in income as a significant resource, and the influence of this on decision-making dynamics within the household, we wanted to understand what influence IDPC might have on other factors which affect the livelihoods of the women and their children.

Matsayi mace is an important aspect in building positive gender relations and maintaining the respect (*mutunci*) and dignity (*daraja*) of women. *Mutunci* is considered a significant resource for women, on which their self-esteem and dignity depend, conveying a force of moral authority in certain situations, and which is controlled by the women themselves. It is defined in relation to the woman's situation and the significant resources that women control – that is, (1) 'position post', meaning the household, with economic activities for women being undertaken within the home; (2) status, often acquired through participation and contribution to the wider family; and (3) the woman's proper place, meaning shielded away from the 'evil eyes of the male'. For this reason, it is strongly linked to the separation of working and economic spheres of men and women, through the seclusion of women (*auren kulle*). Any impact on gender relations, and empowerment of women, would have to be understood in relation to *matsayi mace*.

The impact of IDPC on gender relations was linked to the reinforcement of the seclusion and separation of economic spheres. By contributing to the family's welfare, through petty trading within the confines of the home, women felt that they could gain the respect associated with *matsayi mace*. Conflicts often arise within households when the man's and woman's inability to provide a sufficient livelihood forces the woman out of seclusion to work on fields, which exposes her to community shame. While this shame is often moulded by the tenets of Islam, seclusion has also become an economic symbol of wealth. Women often ask for divorce on the grounds of poverty, and inability to maintain their economic activities in a separate sphere. Poorer households exercise partial seclusion, and a progression towards full seclusion and the maintenance of *matsayi mace* was a major indicator of the impact of IDPC.

Mutunci, or the respect acquired by a woman who is able to maintain her home while in seclusion, was seen as a significant resource for women, as the resulting feeling of dignity, *daraja*, was an important tool in influencing decision making within the marriage, family, and wider community. From this perspective, the wives of adopters felt that the technology had contributed to the welfare of women, particularly towards positive gender relations.

> *Very few women in this area, who are not in full seclusion, can admit it. A woman's place is in her home. She is safe there and will not put herself at risk by affecting her own respectability. She can contribute to her family, and no one will know that it is the wife who provides more. There are of course women who are not happy within their husbands' home, and if their parents are dead, they may choose to remain, and not ask for divorce. IDPC has given women something that they can control and use to influence decision making. And that is the sense of respectability that is associated with seclusion. Not everything is based on money.* (Mariya Liman, Fako, Minjibir Local Government Authority, Kano)

> *There are of course some who want to maintain their respectability, but are still fighting for personal autonomy. For them, the increase in income through adoption of technologies such as IDPC may assist them in their quest, simply because they can choose to move away from their village.* (Alima Tbello, zonal extension agent, Kano National Agricultural Research Agency)

Conclusions

IDPC was developed to address problems of low productivity. Increased productivity has increased the availability of food, fodder, and income. The impact of IDPC on intra-household welfare is linked to the income thus placed in the women's hands, from cash for household goods (*Kudin cefane*) to petty trading of bean cakes and other goods. Surplus income has been found to be extremely important in providing food and nutritional benefits to the home, particularly during periods of risk. Most importantly, income generated through the adoption of IDPC has entered a largely female domain, where transfers of gifts and income reserves are passed on from generation to generation. This has helped to improve the social and economic development for women.

The study of IDPC has provided insights into the community that should be considered when disseminating new technologies, or in encouraging the wider participation of women in agriculture. The 'honeycomb trading' may provide a meaningful avenue through which information on agriculture could be passed on. The use of women's credit groups could provide an interesting avenue for raising or holding cash, in the form of grain/seed reserves.

A re-interpretation of the social or religious 'constraints' could explore provisions existing within the local rubric that allow wider participation of women. The seclusion system can be understood either as a status symbol, or as a liability for women's control of and access to other resources such as health care. Future IITA technologies should go beyond maintaining women's seclusion and the separation of work spheres between women and men. This culture does provide opportunities where technologies could actively engage women in agriculture. Seclusion of the heart (*kulle zuci*) could provide such an opportunity, in which women are able to participate without restrictions on personal autonomy, while 'behaving in a manner that is befitting their status' (Renne 2004).

Acknowledgements

We are very grateful to Adetunji Olanrewaju of IITA and Halima Ahmed of Kano National Agricultural Research and Development Agency (KNARDA) for their research assistance.

Note

1. LPLM – low human-population density (usually defined as fewer than 150 people per square kilometre due to the particularly high population density found throughout Nigeria) and low market access (lack of year-round road access to a wholesale market) (Kristjanson *et al.* 2005); LPHM – low human-population density and high market access; HPLM – high human-population density and low market access; HPHM – high human-population density and high market access. Accordingly, the study areas were stratified into these four resource-use domains, which share similar opportunities and constraints.

References

Adato, M. and R. Meinzen-Dick (2002) 'Assessing the impact of agricultural research on poverty and livelihoods', FCND Discussion Paper No. 128/EPDT Discussion Paper 89, Washington, DC: International Food Policy Research Institute.

Coles, C. (1990) 'The older woman in Hausa society: power and authority in urban Nigeria', in J. Sokolovsky (ed.) *The Cultural Context of Aging*, Worldwide Perspectives, Westport, CT: Bergin & Garvey.

Dey, J. (1992) 'Gender asymmetries in intra-household resources allocation in sub-Saharan Africa: policy implications for land and labor productivity', in *Understanding How Resources Are Allocated Within Households*, Policy Brief 8, Washington, DC: International Food Policy Research Institute and World Bank.

Haddad, Lawrence, John Hoddinott, and Harold Alderman (1994) 'Intrahousehold Resource Allocation: An Overview', Policy Research Working Paper Series 1255, Washington, DC: World Bank.

Hill, P. (1969) 'Hidden trade in Hausaland', *Man* 4: 392–409.

Kibria, N. (1995) 'Culture, social class and income control in the lives of women garment workers in Bangladesh', *Gender and Society* 9 (3): 289–309.

Kristjanson, P., I. Okike, S. Tarawali, B. B. Singh, and V. M. Manyong (2005) 'Farmers' perceptions of benefits and factors affecting the adoption of improved dual-purpose cowpea in the dry savannas of Nigeria', *Agricultural Economics* 32 (2): 195–210.

Manyong, V. M., J. Smith, G. K. Weber, S. S. Jagtap, and B. Oyewole (1996) 'Macrocharacterization of Agricultural Systems in West Africa: An Overview', Resource and Crop Management Monograph No. 21, Ibadan: International Institute of Tropical Agriculture (IITA).

Naved, R. T. (2000) 'Intra-household Transfer of Improved Agricultural Technology; A Gender Perspective', FCND Discussion Paper No. 85, Washington, DC: International Food Policy Research Institute.

Quisumbing, A. R. and J. A. Maluccio (2000) 'Intra-household Allocation of Gender Relations: New Empirical Evidence from Four Developing Countries', Washington, DC: International Food Policy Research Institute.

Renne, E. P. (1996) 'Perceptions of population policy, development, and family planning programs in northern Nigeria', *Studies in Family Planning* 27 (3): 127–36.

Renne, Elisha (2004) 'Gender roles and women's status: what they mean to Hausa Muslim women in northern Nigeria', in S. Szreter, A. Dharmalingam, and H. Sholkamy (eds.) *Qualitative Demography: Categories and Contexts in Population Studies*, Oxford: Oxford University Press.

Schildkrout, E. (1986) 'Widows in Hausa society: ritual phase or social status?' in B. Potash (ed.) *Widows in African Societies: Choices and Constraints*, Stanford, CA: Stanford University Press.

Strategies for out-scaling participatory research approaches for sustaining agricultural research impacts

Aden A. Aw-Hassan

The popularity of participatory research approaches is largely driven by the expected benefits from bridging the gap between formal agricultural science institutions and local farm communities, making agricultural research more relevant and effective. There is, however, no certainty that this approach, which has been mainly project-based, will succeed in transforming agricultural research in developing countries towards more client-responsive, impact-oriented institutions. Research managers must consider appropriate strategies for such an institutional transformation, including: (1) careful planning of social processes and interactions among different players, and documenting how that might have brought about success or failure; (2) clear objectives, which influence the participation methods used; (3) clear impact pathway and impact hypotheses at the outset, specifying expected outputs, outcomes, impacts, and beneficiaries; (4) willingness to adopt institutional learning, where existing culture and practices can be changed; and (5) long-term funding commitment to sustain the learning and change process.

Introduction

Since the early 1990s, participatory research has become an attractive mechanism for conducting adaptive agricultural research. This is mainly motivated by the perception that closer association with resource-poor farmers in identifying problems and involving them in the design and implementation of research offers greater chances of success and adoption of research outputs, thereby enhancing the impact of agricultural research. The advantage of participatory research is considered more prominent in, although not limited to, the adaptation of technologies that require local knowledge of the social, economic, and biophysical environments, or those that need a high level of human capital or require co-operation of different stakeholders. Resource-poor households in dry and marginalised areas who face complex biophysical and socio-economic constraints have benefited less from the successes of agricultural research – for example, those that led to the Green Revolution – than households in well-endowed environments. The result is high prevalence of poverty and malnutrition in the dry areas.

However, there is potential to improve the welfare of these households through agricultural research and development. The development of participatory research and its application came out of the necessity to reach these resource-poor farmers, whose participation in the research and development process is considered essential to bring about desirable changes in rural livelihoods.

The application of participatory research is now well advanced. It has become a resource-mobilising force for project funding. Many research and donor organisations are increasingly embracing this approach. The Consultative Group on International Agricultural Research (CGIAR) has established a system-wide programme to co-ordinate and spearhead participatory research and gender analysis activities (the PRGA Program). The Program runs two community-of-practice forums – participatory plant breeding and participatory natural-resources management. Many methodological guidelines are available. The literature on the subject has also increased substantially. Numerous books and periodicals are produced. There are networks and interactive forums where members exchange ideas and information on-line. Information on the impacts of specific projects is also made available for sharing the lessons learned from experience.

These are significant developments. However, important questions remain: despite these achievements, there is no evidence that the participatory-research movement has yet succeeded in significantly transforming the ways in which agricultural research and development are organised and implemented in dry and marginal areas. In other words, participatory research and gender analysis have not yet been mainstreamed in the agricultural research systems in those areas. This article analyses the factors that influence institutional changes that are favourable to out-scaling participatory research. Here, 'out-scaling' is defined as the large-scale replication and adaptation of successful or promising experiences of the learning process that participatory research fosters. This large-scale dissemination and application of the learning experiences (not the research-generated technological innovations *per se*) is essential for wide-scale impact. The article first highlights some of the essential elements that can facilitate out-scaling of the institutional learning process of participatory research approaches and that could increase the chances of success for agricultural research to have impact. Then follows a discussion of specific strategies that enable participatory research to yield a dynamic learning process, bringing about a profound institutional change which ensures the out-scaling of the benefits of participatory research approaches to larger numbers of beneficiaries.

Process matters

The processes by which innovations are generated, adapted, and disseminated are complex. They involve the direct and indirect interaction of many different entities and organisations among themselves and with end-users. Those who promote the learning-process approach in understanding the spread and impacts of agricultural research argue that the economic model of impact evaluation often treats the process as a black box (Mackay and Horton 2003). The process affects the perceptions, objectives, and activities of different stakeholders, and how information from participatory research is generated, interpreted, and disseminated. They argue that impact evaluation cannot be detached from the process of managing research and generating outputs, which ultimately shapes the development impacts of agricultural research. They point out that evaluation needs not only to provide an account of success, but to explain how and why both successes and failures have been achieved (Mackay and Horton 2003). There is always the desire to generate a blueprint of a model of this process which can be mass-adapted – but this is not likely to happen. There is no ideal model of how to manage complex human processes such as those in agricultural research and development (R&D)

that will fit all socio-economic, political, and cultural situations. But there are important principles that apply in various situations and can provide useful lessons. Hall *et al.* (2003) provide useful suggestions for process-oriented programmes, including the following:

- shift to an innovation systems approach, because the emphasis has to move from a problem-solving framework to a learning framework;
- shift to action-research protocols rather than project-cycle management tools;
- develop projects that involve groupings of local partners (coalitions), where identifying partners becomes part of the research task;
- use stakeholder analysis to make agendas transparent;
- monitor partners' and stakeholders' roles and interests to maintain a poverty focus.

The main point here is that *process matters*. Research institutions are now documenting the processes and institutional innovations developed and used by specific projects, and the wider lesson for other programmes (for example, Shambu *et al.* 2006). Research projects aiming to scale out participatory methods need to closely monitor and document the social processes of interactions among different stakeholders, through self-reflection, and continuously draw lessons from the learning cycles of the project. This allows the development of plausible linkages between impacts and activities, and could facilitate out-scaling of the learning process.

Clarity in programme purpose

In order to understand how participatory research can be out-scaled and institutionalised, it is important first to clarify the purpose of participatory research. The use of such research is often argued to have two main objectives, namely, efficiency (a functional objective) and empowerment (a capacity-building objective). In the former, participation is employed to make use of local knowledge, to better understand farmers' needs, and to improve the effectiveness and efficiency of formal research (Probst *et al.* 2000). In this case, participation is expected to increase research impact by improving the relevance of the technology for users, reducing the research lag (development phase), shortening the adoption lag (early adoption), and increasing adoption speed.

The empowerment or capacity-building objective considers participation to be a means of enhancing local people's capacity for self-directed innovation development (Probst *et al.* 2000). In this case, participation is used as a way of fostering learning through practical experience and working together in all stages of the research process (design, planning, implementation, analysis, and interpretation of results). This is considered a two-way learning process and leads to personal and professional development among local people and researchers. The expected outcomes of this objective are changes in attitudes, improved communication skills, better management skills, and increased organisational capacity. In addition to gaining knowledge and skills through a learning process, this objective suggests that participation is a way of enhancing social change and equity through increased capacity in articulation and negotiation, leadership, and collective action, as well as critical consciousness, and self-esteem among (marginalised) social groups (Probst *et al.* 2000).

These two objectives, although not mutually exclusive, require different levels of intensity in participation and different kinds of skill on the part of the change agents (researchers, extension agents, NGOs, and others). For instance, a participatory approach with a functional objective will require a relatively modest level of participation, which allows proper solicitation of farmers' views and their participation in the assessment of the innovations presented. There are numerous tools available in the literature, both in plant breeding and in natural-resources management, on how to do this effectively (for example, via the PRGA Program website www.prgaprogram.org, retrieved 7 October 2005). It can be done, for example, through participatory technology-evaluation activities such as farmers' crop-variety evaluation and selection. But capacity-building and empowerment objectives aim to support farmers to build their own capacities in collectively managing natural resources, conflict resolution, negotiation, enterprise development, and articulation and communication of their concerns and needs to authorities. The capacity-building objective also involves supporting farmers in efficiently designing their own experiments and recording observations of processes and practices. They can then make decisions based on their own evaluation of new innovations. This requires a different level of involvement and different capacities from researchers. It requires commitment and skills beyond tools for technology assessment. It requires skills such as proper assessment of people's views, perceptions, prejudices, and evaluation of positive and negative feedback. These objectives indicate that it is critical to clarify what kinds of objective are set for a participatory research programme, as these will determine the outcomes and impacts that are to be expected.

Participatory research is now moving towards a learning framework, where the empowerment of users and capacity development – as well as the institutional learning through iterative processes of reflection and subsequent actions – are considered critical intermediate outputs for achieving impacts on the development goals of poverty alleviation and environmental sustainability. This new approach needs a different strategy in impact assessment. The cost–benefit analyses used in economic evaluations of research need to be complemented with a process-based analysis which establishes the linkages between programme activities and outcomes, and which identifies the factors that lead to success. The lessons learned from the latter analysis will facilitate out-scaling of the learning experiences of participatory research.

Clarity in expected outcomes

Participatory research certainly calls for additional resources to accommodate the intensive interaction with farmers. It involves greater co-ordination with different disciplines and stakeholders, such as men and women farmers, farmer-group leaders, extension staff, NGOs, government departments, and development projects. The interaction of these different partners is important, because they all share the same goal of improving rural welfare and they have different strengths and resources. So, with effective co-ordination, these partners can complement each other and enable larger numbers of people to benefit from participatory research. Building such effective partnerships will increase the transaction costs, because of the increase in meetings and workshops to jointly develop common understanding of concepts and objectives, and to formulate programmes of work among partners. All these mean that participatory research is more resource-intensive than conventional approaches. But whether participatory programmes have been cost-effective and whether they have greater benefits that justify these costs depends on the clarity of the objectives of the programme.

The goal of any agricultural research programme in a developing country is to increase food supply and improve the welfare of rural people through improvements in their production systems, measured by specific performance indicators, such as productivity, profitability, reduced drudgery, reduced exposure to chemical pesticides, increased efficiency of natural-resource use, increased awareness, increased skills in management of natural resources, increased entrepreneurial skills, and improved access to markets and sources of knowledge. Depending on the type of participatory research – functional or empowering (capacity building) – it is essential to have clear and specific programme objectives. But because the purpose of participatory research can be interpreted differently by different stakeholders, the expected outcomes can also be interpreted differently. This makes impact assessment difficult. Lilja and Ashby (2001) point out that the great diversity of objectives and expected impacts attributed to participation make it difficult to identify the most important impacts for assessment in relation to different stakeholders' interests.

Difficulties in assessing the impact of participatory research can be a constraint to mainstreaming it as a standard practice of agricultural research organisations, which is essential to scaling it out to larger numbers of beneficiaries over a larger geographical area. Another cost-related argument is that because the innovations generated by participatory research can be considered specific to the conditions under which they are developed, and because of the high degree of heterogeneity of resource-poor households and their farming systems, technologies generated by participatory methods can have only limited impacts. This aspect raises questions of research efficiency. In other words, if participatory research is essential to develop technologies for complex systems where the understanding of the system is critical, how can it have large-scale impact? This implies that each sub-system within a farming system could potentially require the same level of intensive engagement with a variety of stakeholders in order to develop useful innovations. In this case, systems – as defined by market access and agro-ecological and social factors – can vary within short distances. Another critical question is how a research programme which serves larger numbers of beneficiaries, for example in a country, could be designed if it had to adopt farmer-empowering participatory approaches.

Conroy and Sutherland (2004) demonstrate that, with certain conditions and through proper use of the recommendation-domain concept, the potential number of resource-poor beneficiary households could be large. From another perspective, the main appeal of participatory research is its ability to cater for the heterogeneity of resource-poor farmers. Here again the objective of the participatory research is critical. It is stressed that the objective of adaptation trials, for example, is not as much taking measurements as it is about farmers' opinions of the practicality of the idea, with the most important result being the general progress of a group's learning (Petheram 2000). Clearly stated objectives of a participatory research programme are, therefore, essential to assess its cost effectiveness and impacts. This helps to identify which impacts to assess, and will facilitate the institutionalisation and out-scaling of participatory research.

Develop project-impact pathway

The process through which agricultural research has an impact on the livelihoods of rural societies and the environment is influenced by many factors, both internal (at the household level) and external (at the community, regional, national, and international levels). In addition, rural societies are engaged in a dynamic learning process, bring ideas and innovations from a variety of sources, and are influenced by many different development actors (for example, researchers, extension agents, other development agents, traders, and other farmers). Given these dynamic and multi-agent processes, it can be difficult to track the impacts of complex agricultural research outputs (such as natural-resource management technologies) to one specific research project. The benefit–cost model of economic evaluation of research impacts does not provide much insight into how success was achieved. In order to complement the economic model of impact evaluation, Douthwaite et al. (2003) propose the adoption of a two-step impact-evaluation procedure. The first step is that a research project develops an impact pathway which is 'an explicit theory or model which explains how the project sees itself achieving impact'. 'Impact pathway' is defined as a general approach for conceptualising impact processes which provides a framework for research planning and the design of evaluation studies (Springer-Heinze et al. 2003). This can be presented in the form of a flow chart of activities, outputs, effects, intermediate outcomes, and impacts, which specifies a chain of assumptions linking resources, activities, intermediate outcomes, and the programme's ultimate goal (Bickman 1983, cited in Mackay and Horton 2003). The project then uses the impact pathway to guide a self-evaluation to establish the direct benefits in its pilot sites. The second step is that *ex post* impact evaluation of the project is conducted, using the benefit–cost analysis. The use of impact pathway at the outset of the project helps to focus its activities on potential successes, and the self-evaluation provides an opportunity to revise the assumptions in the impact pathway and learn from the process.

Adopt an institutional learning framework

It is now commonly agreed that participatory research involves some kind of a learning process whereby farmers, researchers, and other actors in the innovation network learn from the experiences and knowledge of each other. This process requires that the institutions implementing participatory research also adopt an organisational learning framework. This is defined by Dixon (1994, cited in Conroy and Sutherland 2004) as 'a learning process at all levels of the organisation (individual, group or system level) that allows a continuous transformation of the organisation in a direction that is increasingly satisfying to its stakeholders'.

The contrast to this is a top–down approach, which is prevalent in many agricultural research institutes. This discourages openness and inhibits innovation. It also lacks the individual incentives for taking risks in research. Participants of a facilitation workshop in Punjab Province in Pakistan considered that the top–down culture that was prevalent throughout most of the project components was a hindrance to the success of participation. This organisational learning process requires change in attitudes and practices at the organisation level beyond one specific project. In other words, institutional learning involves 'the process through which new ways of working emerge' (Hall *et al.* 2003). There are, however, examples where an organisational learning process is applied in national research systems in developing countries. One such case, attempting to apply an integrated natural-resource management research approach with focus on improving the livelihoods of African highland dwellers, using participatory research and institutional learning processes, is described in Amede *et al.* (2006). However, this approach is not common among agricultural research institutions. This gap in institutional development towards an organisational-learning framework contributes to the slow institutionalisation of participatory research and its out-scaling.

Although research programmes recognise the importance of research processes and organisational impacts, they often do not consider them as one of their main outputs, but rather a secondary product. This can be partly explained by the long-term nature of assessing the organisational impacts, if any, of a programme. In at least two experiences of the International Center for Agricultural Research in the Dry Areas (ICARDA) – one in Syria and the other in Pakistan – workshop participants, when asked how participatory research can be mainstreamed, said that organisational support was the most important factor, followed by increasing staff capacity and knowledge of participatory methods. An ICARDA study of the capacity needs for institutionalising participatory research and gender analysis documented a number of constraints, including: (1) the top–down culture in the region; (2) limited knowledge about participatory approaches among managers, researchers, and extension staff; (3) lack of interest or resistance among researchers and research assistants; (4) a transfer-of-technology culture; (5) compartmentalised organisational structures; and (6) lack of or weak NGOs and civil societies (Braun 2005).

In other regions, this hierarchical system is considered as a factor that inhibits learning processes and slows down positive organisational change. In reviewing the development of Indian agricultural research and extension, Raina (2003) concludes that 'the hierarchy from policies down to research and from there to extension is an important institutional impediment in achieving the desired impact of agricultural technology'. The hierarchical (top–down) culture is further reinforced by decision makers who aspire to centrally planned economic policies. Agricultural policies and general policy attitudes also play a role in the development of participatory research methods. Policies that mandate production quotas on certain 'strategic' crops prevail in a number of countries in Central and Western Asia and North Africa. These policies discourage farmers from choosing which crops to plant, but support crops that are deemed strategic for food security, employment, and income. But these latter crops may involve practices that are not environmentally sustainable (for example, depletion of ground water, salinisation of land) or may not be economically efficient. Agricultural research and development systems under such a policy environment are unlikely to embrace research approaches which treat farmers as partners in a learning process.

However, even under such circumstances, project-based participatory research activities are being used as important intuitional learning processes. There are many examples of such projects managed by the CGIAR Centres in different countries that enjoy strong political support from national programmes.[1] These projects provide strong learning opportunities with potential impact on organisational and policy changes. However, institutional learning should be deliberate and planned, based on the overriding principle of expanding the impact of agricultural research to the resource-poor farmers. These and future projects need to think about the process much more seriously. The term 'institutional learning and change' is used to describe the process through which institutions actively engage in a learning process through which new ways of working emerge (Hall *et al.* 2003). This learning can be achieved if there is a platform where different actors (different organisations, farmers, researchers, extensionists) can openly discuss and attempt to change the discrepancies in perspectives, objectives, expectations, and organisational features, and rules that constrain the innovation system. Specifically, three ICARDA projects have attempted to apply this principle: the Khanasser valley in north-east Syria, funded by the German Bundesministerium für Wirtschafliche Zusammenarbeit; the Barani Village Development Project, in the dry areas of Punjab province, Pakistan, funded by the International Fund for Agricultural Development (IFAD); and an integrated project in the Karkheh River Basin in Iran, funded under the Water Challenge Program co-ordinated by the International Water Management Institute. Although these and other projects have taken this first step in the right direction, their impacts on institutional learning for ICARDA and the national programmes involved have yet to be evaluated.

Build local partnerships

The adaptation and uptake of innovations for improving the livelihoods of the poor in the dry areas and their environments is a complex process which involves many researchers from different organisations, including international, regional, and national research organisations, development organisations at different levels (local, regional, national, and international), various government and non-government bodies, NGOs, agricultural producers, market agents, and consumers, which form the innovation systems (Horton and Mackay 2003). Building of local partnerships among these stakeholders is essential. This helps in understanding and making use of the local innovation system, in designing knowledge-sharing mechanisms, as well as unlocking potential constraints that may be critical for the process of generating and disseminating technology.

The success of participation can be measured by the overall impact on the welfare of the resource-poor farmers who are not well served by conventional research approaches. In order to be cost-effective, the outreach of the approach should be scaled out to a large number of resource-poor households. There is a need not merely to scale out the functional or efficiency outcomes of participation, but also the capacity-building or empowerment outcomes of participatory research. This clearly is a task beyond the capacity of most research organisations. In an ideal world, where effective extension services exist and civil-society organisations (CSOs) operate, there could be clear division of labour, where researchers focus on the development of innovations, and extension agencies and CSOs would focus on out-scaling and capacity-building aspects of participation. This can be done seamlessly, as one programme, without separation between partners. It requires mechanisms for facilitating multi-stakeholder processes in order to ensure co-ordinated decision making and the smooth flow of information and exchange of knowledge. The division of labour is essential for assigning tasks and responsibilities, harnessing the different strengths of the partners participating in the programme, and harnessing the power of disciplinary focus in tackling critical problems. But the real world is far from this ideal scenario. Weak extension systems and lack of CSOs are the norm rather than the exception in the dry areas. This limits the extent to which participatory research can be scaled out nationally or regionally. The paucity of civil societies and weakness of extension services in many national agricultural systems in Central and Western Asia and North Africa is an obstacle to out-scaling and institutionalising participatory research. A number of project-based experiences of participatory research have been successful in mobilising resources and have shown good results in capacity building and applying the principles of participation.[2] The lesson learned from these projects was that building local partnerships was the most difficult and time-consuming activity. In some cases, weak extension and the absence of a vibrant civil society remain a major critical institutional gap which affects the institutionalisation and out-scaling of participatory research. Nonetheless, farmer groups and other willing stakeholders, including development projects and departments of agriculture, can participate in local partnerships. In Central Asia extension systems are the weakest, and CSOs, with few exceptions, have yet to emerge. As a result, the application of participatory approach in agricultural research and development is still project-based and the experiences are limited.

A critical consideration for research-project planning aiming to apply participatory research is the inclusion of extension services in the programme. This would include activities in capacity building for extension staff and in utilising their human resources in expanding the research and development programme to a larger number of beneficiaries. Although extension services may be weak in many developing countries, the available human resources are often under-used. Thus, effective involvement of extension services would increase the chances of out-scaling the application of participatory research principles and accelerate the transfer of its learning experience.

Focus on stakeholder capacity building

Human capacity is key to the institutionalisation and scaling out of participatory research approaches. There is wide variation in skills and capacities among national programmes in developing counties, but the regional capacity in the dry areas can be generally described as low, because of the very low rate at which participatory research has spread, compared with more favourable areas. The competences that are lacking include facilitation of multi-stakeholder processes, stakeholder analysis, process management, participatory priority setting, impact assessment, and gender-focused analysis. Capacity building should still be a priority in promoting participatory research, impact assessment, and gender analysis. This can be carried out through formal training, combined with on-the-job training and informal mentoring in project settings.

However, lack of follow-up can make formal capacity building ineffective. Weak follow-up of training has been perceived as a concern in a recent survey conducted at ICARDA (Braun 2005). Without follow-up it will be difficult for participants to translate what they have learned into gender-sensitive, participatory research processes. Often lack of adequate funding is cited as the reason for this lack of follow-up. The need for follow-up has to be acknowledged from the beginning – that is, in the project design – to ensure funding. The ICARDA survey also illustrated that capacity building should be more practice-oriented. It is critical that project plans include a programmed and iterative local approach with training and mentoring. This allows opportunities for discussion and reflection on experiences as a frequent and regular element of the process. Another way of supporting capacity building is to organise experience-sharing workshops among practitioners in different projects.

Strengthen co-ordination among stakeholders

When reasonable numbers and different types of organisations are involved in a project that uses participatory methods, co-ordination may be problematic. This is certainly not unique to participatory research: in development in general, Lloyd-Laney *et al.* (2003) point out that better co-ordination of information generation and dissemination is being recognised at every level as necessary to achieving development objectives. Because modern understandings of participatory research encompass multiple objectives, which call for the involvement of a number of stakeholders, co-ordination becomes all the more critical.

There are numerous inter- and intra-institutional barriers to improved co-ordination, as was found in a project in the dry areas of Punjab Province in Pakistan (von Korff 2002), in which the task of co-ordinating among provincial research institutes, national research institutes, the project's development components (including those forming and supporting community organisations and providing micro-finance to women), and ministry departments – who all had a stake in the project – was not easy. Participants at a facilitation workshop identified effective co-ordination of this multi-stakeholder process as critical to the success of the project. In the same workshop, participants recognised the need for their attendance at the monthly community-organisation meetings, where community-development priorities were discussed. A survey of researchers at ICARDA and NARS partners involved in various adaptive research projects revealed that many researchers would like to see more integration occurring across disciplines and with other actors, including national programmes, NGOs, and the private sector. Better co-ordination means greater efficiency in project implementation and ensures that necessary information reaches the target beneficiaries (von Korff 2003). This affects the impact of projects. However, effective co-ordination requires that partners have skills such as facilitation of multi-stakeholder processes. Projects involved in participatory research with multiple partners need to recognise this from the outset and include it in the implementation plan, with adequate allocation of resources.

Consider the development context

Farmers in the dry areas face both environmental and socio-economic conditions that generate a relatively high incidence of poverty. There are, however, many promising technological options available. But in most cases, simply selecting these options is not enough to ensure that they will be adopted. Out-scaling the impact of these technological options requires functioning institutions, so the development of local institutions, rural infrastructure, enterprise linkages, micro-finance, insurance, farmer-interest groups, and other institutional innovations is essential for scaling out the diffusion of these promising technologies.

The main appeal of farmer participation in agricultural research and development is its ability to reach resource-poor farmers and mobilise their knowledge, skills, and experiences to enhance their interactions with formal research for solving problems that affect their livelihoods. A characteristic of resource-poor farmers, particularly in the dry areas, is that they live in environments which lack essential services, such as water supply, health services, and access to markets (commodity and input markets, and financial markets, for example). A reality that confronts any participatory-research practitioner is this general under-development, which negatively affects the welfare of resource-poor households as well as determining their immediate priorities. This general lack of infrastructure and services hampers the ability to implement participatory research effectively.

Under these circumstances, farmers may not see agricultural research as their top priority. This was the case for communities in the Yemeni mountains, where it turned out that farmers had agreed to discuss agricultural research problems only after a consensus was reached that their water-shortage problem would be included on the agenda. Water supply, particularly in the dry season, was an acute problem. Hence the community was very keen on water-supply development (personal observation). Participatory research projects in such areas could be more successful if they worked in co-operation with development projects in the area that can address some of these immediate development issues.

Here it is important to note that poor infrastructure and a lack of rural services are a reality in many dry areas and marginalised environments, and something for which practitioners have to develop strategies for successful participatory research. One such strategy is to form alliances with development organisations that can address the immediate needs of the population, while research focuses on the task of generating and adapting agricultural innovations. But to work successfully with development projects calls for the establishment of good partnerships with clear expectations and roles. Again such alliance building requires skills and attitudes that traditional research organisations do not have. The need and commitment to make an impact on resource-poor households through client-responsive and participatory research approaches is creating a demand for such skills. One insight learned from ICARDA's work with IFAD development projects is that the research activity has to be closely related to one of the main agricultural problems being tackled by the development project, so that the knowledge and data that the research generates can be of immediate use to the development project. Although research outputs may take time, some of the insights may be robust enough to shed new light on the problem and suggest a clear direction for the development project. Such a scenario, if realised, can have considerable impact. An important dimension of the development context is the use of market access as a critical entry point for mobilising rural communities and as a pathway for innovations and change. This market-oriented approach may review research priorities and revise established programme mandates.

A note for investors

Funding is a major factor in the spread of participatory methods in agricultural research and development. The rise in the number of projects that include participatory methods is a clear indication of this influence. There are many good examples of successful projects in the dry areas which apply participatory research approaches, and their number is increasing. In the long run, it is plausible to assume that the knowledge and capacities gained from these projects will reach a critical level which will bring about essential organisational changes that are favourable to participatory research and gender analysis. In other words, these project-based participatory research activities will eventually contribute to institutional development towards an impact-oriented, client-responsive agricultural research system. But the rise in the project-based participatory research activities has not so far brought about this organisational change. One main reason for this is that projects are often designed for a short timeframe, and they are problem-oriented. Projects that address organisational change and issues of structure and capacities at the institutional level are rare, although without additional resources that tackle the organisation and capacity-building issues, the effects of short-term project-based funding of participatory research are unlikely to be sustained.

Conclusion

The participatory research movement has made great strides in promoting a more conscious research programme, where the ultimate users of innovations are involved and participate in the development of innovations. There are numerous cases that measure the impacts of participatory research. Whether the participatory research movement will succeed in transforming the agricultural research systems in such a way that the approach is institutionalised and its results are scaled out to larger numbers of beneficiaries has yet to be determined. Some of the factors that influence this wider intuitional impact include adapting an institutional learning framework, attention to learning processes, building local partnerships, adopting a new impact-evaluation strategy (including development of project-impact pathways), and considering building the capacity of partners. Some of the constraints to out-scaling participatory research approaches in agricultural R&D that are raised in this article are similar to the difficulties that constrain the scaling up of community-driven development. These difficulties include high costs, hostile institutional setting, difficulties in co-operation among different stakeholders, and lack of scaling-up logistics such as training of a large number of participants (Binswanger and Aiyar 2003).

Research administrators should consider these strategies in order to promote an environment which is favourable to participatory research and which has lasting impact. They should also seriously consider farmer participation in setting research priorities and revise established programme mandates as necessary. Donors should consider the effects of short-term funding on the implementation of participatory research projects which require substantial amounts of time to be spent, at least in the beginning, on building relationships and partnerships, both between farmers and researchers and among other stakeholders in the process. These considerations can accelerate the use of participatory approaches in agricultural research and development, and increase the chances that agricultural research will have a positive impact on the poor.

Notes

1. Two current initiatives are the Water Challenge Project in Upper Karkheh River in Iran and the Bright Spot project on Salinity Management in the Aral Sea basin by the International Water Management Institute (IWMI), ICARDA, and the International Center for Bio-saline Agriculture (ICBA).
2. An example of these successful cases is participatory barley breeding, which is now adopted by several national programmes. Another example is the application of integrated natural-resource management now applied in several projects in Iran, Pakistan, and Syria.

References

Amede, Tilahum, Laura German, Sheila Rao, Chris Opondo, and Ann Stroud (eds.) (2006) *Proceedings of the African Highlands Initiative Conference, 12–15 October, 2004, Nairobi, Kenya*, Kampala: AHI.

Binswanger, Hans P. and Swaminathan S. Aiyar (2003) 'Scaling Up Community-Driven Development: Theoretical underpinnings and program design implications', *World Bank Policy Research Working Paper* 3039, available at http://ssrn.com/abstract=636401 (retrieved 17 September 2007).

Braun, Ann (2005) 'Assessment of Capacity Development for Participatory Research and Gender Analysis among ICARDA and Partner Institutions', unpublished consultancy report, Aleppo: ICARDA.

Conroy, Czech and Alistair Sutherland (2004) 'Participatory technology development with resource-poor farmers: maximising impact through the use of recommendation domains', *Agricultural Extension and Research Network Paper* No. 133, London: Overseas Development Institute, available at http://www.odi.org.uk/agren/papers/agrenpaper_133.pdf (retrieved 17 September 2007).

Dixon, N.M. (1994) *Organizational Learning: Becoming Intentional*, New York, NY: McGraw-Hill.

Douthwaite, B., T. Kuby, E. van de Fliert, and S. Schulz (2003) 'Impact pathway evaluation: an approach for achieving and attributing impact in complex systems', *Agricultural Systems* 78 (2): 243–65.

Hall, A., V. R. Sulaiman, N. Clark, and B. Yoganand (2003) 'From measuring impact to learning institutional lessons: an innovation systems perspective on improving the management of international agricultural research', *Agricultural Systems* 78 (2): 213–41.

Horton, D. and R. Mackay (2003) 'Using evaluation to enhance institutional learning and change: recent experiences with agricultural research and development', *Agricultural Systems* 78 (2): 127–42.

Lilja, Nina and Jacqueline A. Ashby (2001) 'Overview: assessing the impact of using participatory research and gender/stakeholder analysis', in Nina Lilja, Jacqueline A. Ashby, and Louise Sperling (eds.) *Assessing the Impact of Participatory Research and Gender Analysis*, Cali, Colombia: PRGA Program, available at http://idrinfo.idrc.ca/archive/corpdocs/117290/quitobook.pdf (retrieved 7 October 2005).

Lloyd-Laney, Megan with contributions from Andrew Scott, Heather Mackay, and Rona Wilkinson (2003) 'Making Knowledge Networks Work for the Poor', final report, Rugby: ITDG, available at: http://www.practicalaction.org/docs/consulting/mknwp%20project%20final%20report.pdf (retrieved 14 September 2007).

Mackay, R. and D. Horton (2003) 'Expanding the use of impact assessment and evaluation in agricultural research and development', *Agricultural Systems* 78 (2): 143–65.

Petheram, John (2000) 'A manual of tools for participatory R&D in dryland cropping areas', report for the Rural Industries Research and Development Corporation, University of Melbourne, Creswick, Victoria, Australia.

Probst, Kirsten, Jürgen Hagmann, Thomas Becker, and Maria Fernandez (2000) 'Developing a Framework for Participatory Research Approaches in Risk Prone Diverse Environments', paper presented at Deutscher Tropentag 2000, Hohenheim, 11–12 October.

Raina, R. S. (2003) 'Disciplines, institutions and organizations: impact assessments in context', *Agricultural Systems* 78 (2): 185–211.

Shambu, Prasad C., T. Laxmi, and S. P. Wani (2006) 'Institutional Learning and Change (ILAC) at ICRISAT: A Case Study of the Tata-ICRISAT Project', *Global Theme on Agroecosystems Report* No. 19, Patancheru, Andhra Pradesh, India: ICRISAT.

Springer-Heinze, A., F. Hartwich, J. S. Henderson, D. Horton, and I. Minde (2003) 'Impact pathway analysis: an approach to strengthening the impact orientation of agricultural research', *Agricultural Systems* 78 (2): 267–85.

von Korff, Yorck (2002) 'Report of a Workshop for Members of the Barani Village Development Program in Islamabad on 16–22 September 2002', unpublished consultancy report, Aleppo: ICARDA.

von Korff, Yorck (2003) 'Improving and Institutionalizing Farmer Participatory Research in the KVIRS project', unpublished consultancy report, Aleppo: ICARDA.

Integrating participatory elements into conventional research projects: measuring the costs and benefits

Andreas Neef

Until recently, participatory and conventional approaches to agricultural research have been regarded as more or less antagonistic. This article presents evidence from three sub-projects of a Thai–Vietnamese–German collaborative research programme on 'Sustainable Land Use and Rural Development in Mountainous Regions of Southeast Asia', in which participatory elements were successfully integrated into conventional agricultural research as add-on activities. In all three sub-projects the costs of studying local knowledge or enhancing farmers' experimentation consisted of additional local personnel, opportunity costs of participating farmers' time, and travel costs. However, these participatory elements of the research projects constituted only a small fraction of the total costs. It may be concluded that conventional agricultural research can be complemented by participatory components in a cost-effective way, while producing meaningful benefits in terms of creating synergies by blending scientific and local knowledge, scaling up micro-level data, and highlighting farmers' constraints affecting technology adoption.

Introduction

Assessing or measuring success of agricultural research is increasingly important in order to justify national governments' spending on agricultural research and to raise additional funds from international donors. In the words of Alston *et al.* (1995: 19), 'agricultural research is an investment in the production of knowledge. It competes with other activities for scarce resources.' This applies also to participatory research, which is often regarded by its critics as producing only site-specific results, hence lacking reproducibility and being too time-consuming and costly, particularly in terms of scaling up.

There is an increasing body of literature on assessing costs and benefits or impact of participatory research (see, for example, Johnson *et al.* 2001; Smale *et al.* 2003; Lilja *et al.* 2004). Only a few of them, however, focus on projects in which participatory approaches were applied as add-on activities in conventional agricultural research projects.

Drawing on experience from three sub-projects of a Thai–Vietnamese–German collaborative research programme on 'Sustainable Land Use and Rural Development in Mountainous Regions of Southeast Asia', I compare the costs of conventional research approaches with studies of local knowledge and farmers' experimentation that were added as a complementary participatory component. I argue that many conventional agro-ecological and agronomic studies are rather reductionist, site-specific, and expensive, while low-cost participatory, add-on elements can substantially validate the significance of the research for a greater area and for different agro-ecological and socio-economic conditions.

Costs and benefits of adding a participatory component to conventional research projects: case studies from the Uplands Program

The Uplands Program is a long-term collaborative research programme entitled 'Sustainable Land Use and Rural Development in Mountainous Regions of Southeast Asia'. Its study areas are located in various districts in north Thailand and north-west Vietnam, home to diverse ethnic minority groups, such as the Hmong and Black Lahu. Its 16 sub-projects represent a range of disciplines, including soil science, agro-ecology, horticulture, agricultural economics, and rural sociology. A particular feature of the Uplands Program is an 'umbrella' sub-project entitled 'Participatory Research Approaches', of which the objectives are first to support participatory approaches as a cross-cutting issue in all sub-projects, and second to assess the potential and limits of stakeholder participation in different research contexts and research phases.

Study of pests and beneficial insects in lychee orchards

Since the mid-1980s, lychee production has been a major source of income for Hmong farmers in the Mae Sa Noi sub-catchment, which is located about 35 km north-west of the northern Thai capital Chiang Mai. Since the mid-1990s, farmers have incurred significant losses from insect pests and started to apply high doses of broad-spectrum, hazardous insecticides (although little is known about whether these measures can effectively control these pests). In order to increase the knowledge of the major pests and their natural enemies, and to identify more sustainable pest-management strategies, one sub-project of the Uplands Program was set up for a period of about three years (September 2000 to July 2003) to study the complex interaction between pests and beneficial insects in lychee orchards.

Objectives of the study and methods applied For the development of alternative strategies of pest control, we first needed to know which are the major pests, and the degree of damage that they cause. In addition, the major natural enemies of these pests needed to be identified. Therefore, one objective of the study was to analyse the arthropod community of the lychee orchards, to identify the major pest species and their natural enemies, and to record their abundance and occurrence during the different seasons.

The sub-project started with a general collection of agro-ecological data from various farmers' fields. These farmers were neither integrated into the knowledge-generation process, nor were they well informed about the purpose of the project. At the behest of the project leader, the research associate established an on-farm experiment. Negotiations with the field owner led to a research contract in which the rights and duties of both parties were agreed upon. The on-farm experiment was predominantly researcher-controlled: the farmer followed the instructions of the research associate as to the management of the trial. The plot was sub-divided into four equal parts of 0.3 hectares each in order to analyse the influence of different weeding practices (regular weeding versus no weeding at all), in combination with different pest-management strategies (typical pesticide application versus no pesticide application), on the arthropod fauna and tree-undergrowth vegetation in the orchard. A great variety of methods (insect traps, hoovering, intoxication, direct observation) were applied in the experimental orchard to catch pests and beneficial insects that were later identified in the laboratory.

Shortly after the trial was set up, the limitations of this reductionist approach became obvious. It was not possible to determine spatial distribution patterns of insects or their migration behaviour and temporary habitats in different life-stages. Following discussions between leaders of this sub-project and a member of the sub-project 'Participatory Research Approaches', a joint study of local knowledge was initiated, involving a group of farmers interested in the subject. The researchers also provided feedback to farmers on the results of the on-farm experiment (for example, on pest occurrence, types and severity of damage), which gave room for a joint analysis of the findings.

Regular meetings with individual farmers and group discussions were organised to obtain information about pest problems and farmers' strategies to cope with them. Farmers were encouraged to make regular checks in their orchards and bring typical insect species in different life-stages for analysis by the scientists, and for discussion about the reasons for their occurrence and their lifecycles and habitats. One problem with the meetings was the fluctuation in attendance, which ranged from 4 to 18 farmers. Field excursions with farmers and demonstrations of pests and beneficial arthropods were also organised.

Results of the study and benefits of integrating local knowledge of insects One of the results of the 'scientific' part of the study was a detailed knowledge of the major pests invading the experimental plot during the three-year research phase. Another outcome was an understanding of the effect of different combinations of pesticide treatments and undergrowth management on the occurrence of pests and beneficial insects. These results, however, could not be easily extrapolated to the sub-catchment level, because of the lack of spatial data.

In contrast to this reductionist approach, the eliciting of local knowledge and local observation helped to give an idea of the major pest insects in the whole sub-catchment that could be more intensively studied in a follow-on three-year phase of the project (July 2003 to June 2006). The farmers emphasised the importance of studying the interaction between fruit orchards and adjacent forest areas, an approach which has been integrated into the second research phase. One of the participating farmers had developed his own hypothesis on the occurrence of a certain pest. He stated that the Asian ambrosia beetle (*Xylosandrus* sp.) could only be found on those trees in his orchard which he had girdled in order to induce flowering. His theory was that girdling increases the sugar content of the sap, which attracts the beetle. An entomologist confirmed that it is likely that the beetle is found only on girdled trees, although his explanation differed from that of the farmer: he suspected that girdling is physically injuring the tree, which would make it more vulnerable to bark-feeding beetles.

Another important outcome of the interaction between farmers and scientists was that awareness could be raised about the ineffectiveness of certain pest-management strategies against particular insects, such as spraying of pesticide against a scale insect (*Laccifera lacca*) and various bark-feeding insects (for example, *Indarbela dea*, *Xylosandrus* sp.). Overall, the participatory component laid the ground for identifying promising biological and integrated pest-management strategies, which were tested in follow-up on-farm experiments.

The study of local knowledge about insects also gave a more realistic picture of farmers' ecological knowledge. Ethnic-minority farmers in the northern Thai highlands use their intimate knowledge about insects, for example, to determine soil fertility (Tinoco-Ordónez 2003) and to make weather predictions (Choocharoen, personal communication). However, this was not the case in this study in the Mae Sa Noi sub-catchment: since lychee farming is a relatively new practice for Hmong farmers and some of the insect species have invaded the orchards only recently, many farmers have little knowledge about these insects. For many of them no local names existed, and terms for 'larvae' and 'caterpillars', for instance, were used interchangeably. Beneficial insects were practically unknown. Also we could not detect a uniform classification system in the form of a 'Hmong insect taxonomy'; instead, we found that farmers' knowledge and classification skills were fairly heterogeneous. Aside from differences in education and age, one major underlying reason was that many farmers sell the entire harvest to traders at the time of fruit-setting, when the yield can already be roughly estimated. The traders then take over the responsibility for applying fertiliser and pesticides, while the farmer's task is reduced to irrigating the trees during dry periods. Hence, most farmers do not regularly check their orchards for insects, a fact which raises concerns about the future prospects for integrated pest management (IPM), a system which relies strongly on collective action and a functioning monitoring system of pest insects.

Comparing the costs of eliciting local knowledge and generating scientific knowledge of pests and beneficial insects The costs of generating scientific knowledge of pests and beneficial insects are extremely high (Table 1). The study needs to be fairly long-term (at least three years) to avoid seasonal biases. There is a need for specialised staff with good taxonomic knowledge to determine the thousands of insects that are caught each month. Collecting a wide range of insects with different movement patterns (flying at different heights, crawling, etc.) and other behavioural characteristics requires highly sophisticated traps and regular field visits to check the traps. In this particular study, traps were monitored every second day, which was made possible by the involvement of a Thai research assistant and a local Hmong assistant (a relative of the field owner).

As compared with the costs of the 'scientific' study, the local-knowledge study was relatively inexpensive (Table 2). The major part of the costs was additional labour: a Thai–German

Table 1: Costs of the scientific study of pests and beneficial insects

Cost category	Details	Cost in EUR
Personnel (expatriate)	Junior researcher (3 years)	91,075
Personnel (local)	Thai research assistant (2 years), ethnic-minority assistant (2 years)	5,865
Equipment	Diverse insect traps (malaise traps, light traps, etc.), binoculars, laptop	14,670
Consumables	Plastic bottles for collecting insects, nets, small insect traps, chemicals, etc.	5,595
International travel	3 international flights (Germany–Thailand–Germany)	3,080
Local travel	Trips to research plot (3 times weekly)	7,660
Total costs		**127,945**

Source: Author's calculation.

Table 2:   Costs of the study of local insect knowledge

Cost category	Details	Cost in EUR
Personnel (expatriate)	Senior scientist as project adviser (4 days), junior researcher (9.5 days)	2,091
Personnel (local)	Thai facilitator (16 days), interpreter for ethnic-minority language (5.5 days)	1,138
Opportunity costs of time	Farmers' participation in 5 group discussions (evening) and various field visits (daytime)	106
Local travel	20 visits to the village (gasoline, maintenance, etc.)	285
Consumables	Small plastic bottles for collecting insects, pens, markers, paper	25
Total costs		**3,645**

Source: Author's calculation.

facilitator for the meetings and discussions with farmers, a Hmong interpreter for some of the meetings, and advice from an expatriate senior scientist on participatory research methods. The second most important cost factor was travel: several trips to the village were in vain because farmers were not always available, due to unforeseen events such as funerals or even police raids during an anti-drug campaign. Owing to the relatively low household income and the fact that most discussions were held in the evening when farmers had some free time, the opportunity costs of farmers' time were comparably low. Material costs (consumables) were negligible.

Local and scientific soil classification in the Black Lahu village of Bor Krai

In this section, I describe the processes of recording local knowledge of soil characteristics in Bor Krai, a Black Lahu community in Mae Hong Son province, northern Thailand, and blending it with scientific knowledge. The survey was conducted by a multidisciplinary group of scientists, namely from soil science, agricultural extension, farming systems, and rural sociology, during the dry season (October to May) 2004/05.

Objectives of the study and methods applied Detailed soil information on watershed scale remains scant in northern Thailand, particularly on sloping land where traditional soil-mapping approaches are arduous and time-consuming. The major objective of this study was to identify a more efficient way to scale up soil information to the watershed level. One of the study areas was the Black Lahu community of Bor Krai, located in a typical limestone area. Soil samples were collected and then analysed in the laboratory; soil types were determined according to the FAO's World Resource Base (WRB) classification (FAO 1998). About 356 field augerings and 23 soil-profile descriptions were carried out for an area of approximately 8.5 km^2. The scientific soil classification was accompanied by a petrographic mapping exercise (Schuler *et al.* 2006).

Information on local soil classification and spatial soil-type distribution was collected during a joint survey by a soil scientist, a farming systems expert, and a rural sociologist, using a range of participatory methods like group discussions, semi-structured interviews with key informants, and participatory mapping. The first step was to identify farmers with an outstanding knowledge of local soils, based on their practical experience. The scientists and a key informant together drew up a list of 11 male and female farmers whose knowledge of soil properties of the area was said to be above average. These farmers were mainly older people who had lived and worked in the area for several years and owned land on various sites in the village area. Next, these farmers were openly asked in a group discussion about the soil types that they could distinguish, and the parameters for this classification. This approach gave a general overview about the local soil knowledge, and the results were the basis for the following investigation. To obtain a more detailed and specific local soil classification, farmers were also asked individually during field walks about the main soil types and their properties. Each farmer led the research team to specific sites, according to his or her soil classification. On site, the farmers were interviewed about main soil properties like water-infiltration rate, water retention, fertility, and suitability for crops (Schuler *et al.* 2006).

Farmers were also interviewed in their homesteads. During these interviews the collected soil samples were shown, and the farmers were asked to identify and name the soils. In a final group discussion, farmers were presented with all soil samples, which they had to sort and rank according to different parameters to crosscheck the former information and to obtain a group consensus about the classification. Finally, a participatory soil-mapping exercise was carried out. Farmers were asked to identify different soil types according to the local classification at many different sites on a topographic map. To facilitate the farmers' orientation, the map included roads, streams, springs and specific landmarks like rocks and caves. The scientist, who did the scientific soil classification, and his local Lahu assistant had an excellent knowledge of the area, and thus were able to give the farmers sufficient spatial information while elaborating the map. Since the topographic map was geo-referenced, the local soil map could be digitised with ArcView and used for further analysis and comparison in a geographic information system (GIS) (Schuler *et al.* 2006).

Results of the study and benefits of integrating local knowledge Similar to other studies of local soil knowledge (for example, Ettema 1994; Talawar and Rhoades 1998; Barrera-Bassols and Zinck 2003), Lahu farmers in Bor Krai classified soils according to recognisable and easily identifiable properties in the field. Their main criteria for classification were topsoil colour, hardness, consistency, suitability for different crops, and yield potential. During a ranking exercise, it was confirmed that the major criterion for soil classification was colour. Farmers distinguished four main soil types: black, red, orange, and yellow. The farmers were then asked to map these four soil types on a topographic map. With this approach, the local soil map reflected a rather rough distribution of the soil types. Many variations in colour and texture that farmers mentioned during the field survey were neglected during the local mapping process.

The scientific soil classification according to the WRB standard produced eight major soil types, divided into further sub-units. The spatial, patchy patterns of the soil types on the scientific soil map were completely different from the local soil map, which was not surprising given that the WRB classification is based on a mix of physical and chemical soil properties in different horizons (soil layers), in contrast to the farmers' focus on topsoil properties and easily perceptible, morphological soil characteristics.

On the other hand, the local soil map showed striking similarities with the petrographic map (compare Figures 1 and 2). One 'scientific' explanation for this strong correlation was that the iron contents of the parent material have a crucial impact on the subsoil colour (Schuler, personal communication). As most of the soils in sloping areas of northern Thailand show a certain degree of erosion, the petrographic origin of the soils increasingly becomes a determinant of the soil colour and hence correlates strongly with local soil types (Schuler *et al.* 2006). The fact that

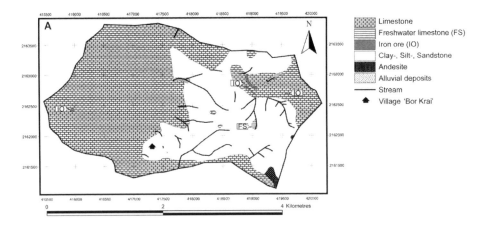

Figure 1: Petrographic map of Bor Krai
Source: Data and map from Schuler, adapted from Schuler *et al.* (2006).

local people attribute certain soil-fertility parameters to soil colour helps to make soil classification more relevant for local land-use decisions.

Another outcome of the local soil-classification process was that farmers' knowledge of the landscape could help the soil scientist to capture the full heterogeneity of the soil units. One soil type (a so-called *Chernozem*) could be detected only with the help of a local farmer, despite being found only a few metres from one of the sampling points for the scientific soil classification.

Overall, these results suggest new research avenues in future phases of the research programme: to integrate local soil mapping and scientific knowledge (petrographic mapping) for scaling up soil information from the field to the watershed and landscape levels.

Comparing the costs of eliciting local soil knowledge and generating scientific soil knowledge Laboratory analyses of soil samples are by far the highest cost factor in the scientific soil classification (Table 3). To distinguish certain soil types (for example, to differentiate between an Acrisol and a Ferralsol), very costly analyses need to be conducted. The labour costs

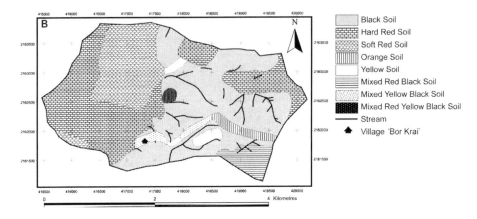

Figure 2: Local soil map of Bor Krai
Source: Data and map from Schuler, adapted from Schuler *et al.* (2006).

Table 3: Costs of the scientific soil classification according to WRB

Cost category	Details	Cost in EUR
Personnel (expatriate)	Junior researcher (3 months and 1 week)	8,505
Personnel (local)	Ethnic-minority research assistant (3 weeks), field workers for digging soil profiles (20 days)	280
Soil analyses	Labour costs, consumables and investment costs for diverse analyses of 8 different soil types (soil texture, water content, organic matter, carbon exchange capacity, base saturation, carbonate content, etc.)	36,945
Equipment/consumables	Augering equipment, plastic bags for soil samples, etc. (shared-cost basis)	150
Travel costs	5 trips to the research area, shipment of samples to Germany for further analysis	435
Total costs		**46,315**

Sources: personal communication by Ulrich Schuler; 'Manual for Soil Analysis of the Department of Soil Science', University of Hohenheim (Herrmann, unpublished); author's calculation.

of the fieldwork are relatively low by comparison, although taking soil samples in this rough terrain is physically extremely challenging. Costs for equipment and consumables were negligible, since they could be also used for studies in other areas.

The petrographic mapping consisted of an additional month's work by the soil scientist with some help from his local assistant (Table 4). The mineralogical analysis in the laboratory amounted to less than one per cent of the costs of the soil analysis.

The local soil mapping was the cheapest part of the whole mapping exercise (Table 5). Aside from the labour input of the soil scientist and his local research assistant, a Thai junior researcher helped in organising and facilitating the individual interviews and group discussions. Minor cost factors were the opportunity costs of farmers' participation and some consumables, such as material for drawing and food for the farmers.

Hence, combining petrographic mapping and local soil mapping can considerably reduce the costs of making fairly accurate maps which might be even more relevant for farmers' land-use decisions and government agencies' land-use planning procedures than maps based on traditional approaches to scientific soil classification.

Table 4: Costs of making a petrographic map

Cost category	Details	Cost in EUR
Personnel (expatriate)	Junior researcher (1 month)	2,460
Personnel (local)	Ethnic-minority research assistant (2 weeks)	40
Mineralogical analysis	Labour costs, consumables, and investment costs for mineralogical analysis	305
Local travel and shipment	2 trips to the research area, shipment of samples to Germany for further analysis	144
Total costs		**2,949**

Source: personal communication by Ulrich Schuler; author's calculation.

Table 5: Costs of preparing a local soil map

Cost category	Details	Cost in EUR
Personnel (expatriate)	Junior researcher (18 days)	1,680
Personnel (local)	Thai junior researcher (1 week), ethnic-minority research assistant (3 weeks)	144
Opportunity costs of time	Farmers' participation in 1 group discussion (evening) and 10 field visits (daytime)	30
Local travel	1 trip to the research area (fuel, maintenance, car rental)	47
Consumables	Pens, markers, paper, colour prints, snacks for meeting with farmers	25
Total costs		**1,926**

Source: personal communication by Ulrich Schuler; author's calculation.

Farmers' experiments with vegetative propagation of Arachis pintoi *as a complement to researcher-managed cover-crop trials*

On the hillsides of northern Thailand, erosion and soil-fertility depletion are among the major constraints to sustainable production of fruit orchards. In a 28-month study, agronomists of the Uplands Program investigated the potential of various cover legume species for erosion control, weed control, and soil improvement, and as forage plants in smallholder hillside orchards.

Objectives of the study and methods applied Two on-station trials and one controlled on-farm experiment (each with three replications) were established at three sites of contrasting altitudes. Cover densities, seed-production potential, and forage quality were assessed using standard agronomic methods.

Hmong villagers of Mae Sa Mai, where the on-farm experiment was conducted, showed particular interest in *Arachis pintoi*, a perennial, stoloniferous legume species. Farmers' experiments with vegetative propagation of *Arachis pintoi* were initiated by a presentation in mid-2003 for about 40 male and female farmers. The participants became actively involved in discussions on potential benefits and risks of growing cover crops in their fruit orchards. The presentation was followed up by a field day, with the participation of seven farmers who were particularly interested in the potential of *Arachis pintoi* as animal feed and ground cover in their fruit orchards. Six farmers finally decided to test the cover legume in their own plots. Researchers gave a short instruction session on the techniques of vegetative propagation and distributed plastic bags containing stolons from research stations (farmers decided how many stolons they required). The experiments were entirely farmer-managed and jointly monitored by farmers and researchers. Farmers decided on location and size of the experimental plots and were responsible for planting and maintenance (weeding, fertilising). Planting densities of 50 × 50 cm were recommended by the researcher, but most farmers varied the planting densities according to their own judgement. The researchers distributed field books to the farmers for regular recording of labour and other inputs. As some of the farmers were illiterate, a research assistant helped in keeping the field book. Cover density of *Arachis* in the experimental plots was monitored by the researchers at monthly intervals during a five-month establishment phase.

In experimenting with *Arachis pintoi* in their hillside orchards, farmers were surprisingly creative. Some created their own experimental designs, such as varying weed management prior to stolon planting and comparing effects of different light intensities. One farmer planted the cover crops in his tangerine orchard in order to increase the soil moisture and reduce evapo-transpiration. The experimental capacities of farmers in these trials confirmed findings of other authors which suggest that farmers' experimentation has a much more formal character than is often expected by scientists, and that combining station-based experiments and farmer-managed trials can provide valuable synergetic effects (see, for example, van Veldhuizen *et al.* 1997; Sumberg *et al.* 2003).

Results of the study and benefits of complementing researcher-controlled trials with local experimentation In the scientific study, *Arachis pintoi* turned out to be one of the best-performing legume species in terms of potential for forming a dense soil cover within several months, particularly at medium and high altitudes. It also showed a high nutritive value with comparatively low fibre content. Hence, the agronomist concluded that it was a promising technology for fruit orchards in the Thai highlands.

Data on the participatory vegetative propagation of *Arachis pintoi* were recorded on experimental plots of five farmers (one member of the initial group abandoned the experiment after a few weeks due to lack of time to maintain the plot). One farmer subdivided his plot so that one part was treated with herbicides prior to planting *Arachis* and one part was hand-weeded before planting. Data on cover densities suggest a high variation of ground cover during the establishment phase of *Arachis pintoi* (Table 6). In general, results under real, i.e. farmer-managed,

Table 6: Comparison of controlled experiments and farmer-managed trials

Degree of participation	Type of experiment	Fertiliser dose	Cover density (months after seeding, in %)				
			1	2	3	4	5
	Station (controlled, 3 rep.)	low	2.8	23.1	57.5	86.7	95.0
	Station (controlled, 3 rep.)	high	3.0	30.9	65.8	85.0	93.3
	Station (controlled, 3 rep.)	low	4.5	47.8	96.7	98.3	100
	Station (controlled, 3 rep.)	high	5.0	56.5	97.5	99.2	100
	On-farm (controlled, 3 rep.)	low	2.8	13.5	55.0	91.7	96.7
	On-farm (controlled, 3 rep.)	high	3.0	16.5	49.2	94.2	95.8
	Type of experiment	Labour input	Cover density (months after transplanting, in %)				
			1	2	3	4	5
	On-farm (farmer-managed)	high	12.0	61.6	70.8	80	60
	On-farm (farmer-managed)	high	12.0	50.4	68	90	70
	On-farm (farmer-managed)	low	6.0	9.4	13.2	13.2	10
	On-farm (farmer-managed)	low	5.2	30.4	19.2	18	16
	On-farm (farmer-managed)	low	10.0	34.8	30.8	30	25
	On-farm (farmer-managed)	low	20.0	43.5	70.4	20[†]	25.3

[†]Effect of herbicide application.
Source: data provided by Schultze-Kraft, adapted from Neef *et al.* (2007).

conditions were far less promising than those from the controlled trials, although the latter were generated from seeds, while the former were from vegetative propagation, which should enable a faster establishment of ground cover. Farmers reported various problems in the first months, such as adverse effects caused by periods of drought, insect pests, and weed competition, the latter particularly after using fertiliser. Sensitivity to drought was indicated by the fact that cover densities dropped slightly in all plots at the beginning of the dry season (December, i.e. Month 5). Lack of tolerance to herbicide use prior to planting and during the establishment phase observed in the farmers' experiments needs further confirmation in controlled experiments. Labour input turned out to be a decisive factor in establishing *Arachis*. Farmers who frequently weeded their plot (see plots with high labour input in Table 6) achieved the highest cover densities of *Arachis*. In the case of two farmers, weeding time interfered with other economic activities, such as growing vegetables and working off-farm.

The major conclusion from the farmer-managed experiments was that farmers had to invest a considerable amount of time in the establishment of a crop whose direct benefits accrue mostly in the long term. This confirmed suggestions of other authors that adoption of multi-purpose cover crops may be a complex and slow process (see, for example, Rivas and Holmann 2000; Wünscher *et al.* 2004). It was concluded that farmers in the study area were unlikely to adopt cover crops on a larger scale. As cover crops like *Arachis pintoi* can provide substantial indirect benefits for downstream users in terms of reducing erosion and preventing landslides, temporary subsidies to support the establishment phase of the crops may be justified. Such measures, however, will depend on political will and/or lowland residents' willingness to pay for such environmental services.

Overall, the benefits of this participatory element did not derive from an increase in adoption rates of better-suited technologies, as is often the expected outcome of participatory research. Rather it helped to avoid future costs of doing research into technologies that are not compatible with farmers' constraints and/or alternative income opportunities (see, for example, Lilja *et al.* 2004, for the case of ICRISAT's mother–baby trials).

Comparing the costs of researcher-managed field trials and farmers' experimentation with cover crops The researcher-managed field experiments with cover crops were less costly than most other 'scientific' studies in our research programme (Table 7). This was mainly due to the fact that the main fieldwork was carried out by a Thai PhD student on a local salary, rather than by an expensive expatriate junior researcher. The costs of a Thai

Table 7: Costs of researcher-managed field trials with 14 cover crops

Cost category	Details	Cost in EUR
Personnel (local)	Thai junior researcher (2.5 years), Thai research assistant (6 months), field workers (weeding, etc.)	13,905
Equipment	Diesel mower, field balance, laboratory freezer, greenhouse	2,015
Consumables	Seeds, fertiliser, herbicides, fences, plastic bags, etc.	2,760
Laboratory analysis	Soil analysis, dry matter, feed-quality parameters, etc.	8,370
Local travel	Trip to research plots (twice weekly; fuel, maintenance, etc.)	3,740
Total costs		**30,790**

Source: Author's calculation.

Table 8: Costs of farmer-managed experimentation with *Arachis pintoi*

Cost category	Details	Cost in EUR
Personnel (expatriate)	Senior scientist as project adviser (1.5 days)	368
Personnel (local)	Thai junior researcher (3.5 days), Thai research assistant (23 days)	266
Opportunity costs of time	Farmers' participation in village meeting, field day and various field visits, labour on experimental plots	212
Local travel	42 visits to the village and experimental plots (fuel, maintenance, car rental)	840
Material	Plastic bags for stolons, field books, fertiliser, food during village meeting and field day	40
Total costs		**1,726**

Source: Author's calculation.

research assistant and of field workers who were employed on a daily basis were also quite low. The costs of the laboratory analyses were comparably high, given that they were conducted at Chiang Mai University, where the costs of hiring technical assistants are rather low, compared with international standards. Equipment and consumables were also a major cost factor. As the Thai PhD student visited three research sites at regular intervals, the travel costs were correspondingly high.

Travel costs were also the major cost factor in the farmer-managed trials with *Arachis pintoi*, followed by labour costs, mainly for a Thai research assistant, who made regular monitoring visits to farmers' fields and helped to keep the field books (Table 8). Opportunity costs were comparatively low, mainly because farmers devoted only a small portion of their fields to the experiments. None of them designated an area for the experiment of more than one per cent of the total field size. The largest size among the experimental plots was less than 0.1 hectares.

Synthesis and conclusions

In all three cases presented above, the participatory component constituted only a fraction of the costs of the conventional agricultural research project (2.8 per cent, 4.2 per cent, and 5.6 per cent, respectively). Major cost factors were additional personnel and travel costs, while opportunity costs of time for participating farmers and costs of consumables were negligible.

Compared with these moderate costs, the benefits from all participatory components were much more substantial. In the first case, 'Pests and beneficial insects', the study of local knowledge of insects contributed to widening the reductionist scope of the 'scientific', conventional research component and to forming a more complete picture of the whole sub-catchment. Together with the conventional research, it also laid the foundation for targeting the major pest insects with more effective and, at the same time, more sustainable pest-management strategies, such as applying bio-insecticides and enhancing the population of natural enemies in the second project phase. In the second study, 'Local and scientific soil classification', researchers identified a new, cost-efficient strategy for scaling up soil information from the field to the landscape and regional levels. In the third research phase (July 2006 to June 2009), it is planned to combine radiometric data for petrographic mapping with recording local soil knowledge. It is expected that this strategy will not only reduce the costs of scaling up, but also make soil maps more relevant for land-use decisions. In the third case, 'Farmers' cover-crop experiments', scientists could identify the major constraints and opportunity costs of farmers participating in the trials. Thus, they could obtain a more realistic assessment of the local adoption potential for a specific cover legume, and the policy changes that are needed to support adoption of soil-conservation measures, if this is a priority of the wider society.

I conclude that participatory approaches can inform conventional agricultural research in a cost-effective way by widening the scope of site-specific experimental set-ups, by supporting the scaling up of micro-level data, and by highlighting farmers' specific constraints in early stages of the innovation process.

Acknowledgements

I would like to thank Ulrich Schuler for providing much of the data needed for the cost comparison in the second case study. Special thanks are extended to Guy Manners for his thorough review of the paper and helpful editorial comments. I am also indebted to the participants of the Impact Assessment Workshop at CIMMYT Headquarters in Mexico in October 2005 for their helpful comments on an earlier version of this paper. The financial support of the Deutsche Forschungsgemeinschaft (German Research Foundation) in carrying out this study is gratefully acknowledged.

References

Alston, Julian M., George W. Norton, and Philip G. Pardey (1995) *Science Under Scarcity: Principles and Practice for Agricultural Research Evaluation and Priority Setting*, Ithaca, NY and London: Cornell University Press.

Barrera-Bassols, Narciso and Albert J. Zinck (2003) 'Ethnopedology: a worldwide view on the soil knowledge of local people', *Geoderma* 111: 171–95.

Ettema, Christien H. (1994) *Indigenous Soil Classifications. What Are Their Structure and Function and How Do They Compare with Scientific Soil Classifications*, Athens, GA: Institute of Ecology, University of Georgia.

FAO (Food and Agriculture Organization of the United Nations) (1998) *World Reference Base for Soil Resources*, Rome: FAO.

Johnson, Nancy, Nina Lilja, and Jacqueline Ashby (2001) *Characterizing and Measuring the Effects of Incorporating Stakeholder Participation in Natural Resource Management Research: Analysis of Research Benefits and Costs in Three Case Studies*, Working Document No. 17, Cali, Colombia: CGIAR Systemwide Program on Participatory Research and Gender Analysis.

Lilja, Nina, Jacqueline Ashby, and Nancy Johnson (2004) 'Scaling up and out the impact of agricultural research with farmer participatory research', in Douglas Pachico and Sam Fujisaka (eds.) *Scaling Up and Out: Achieving Widespread Impact Through Agricultural Research*, Cali, Colombia: Centro Internaçional de Agricultura Tropical (CIAT).

Neef, Andreas, Rupert Friederichsen, Benchaphun Ekasingh, Dieter Neubert, Franz Heidhues, and Nguyen The Dang (2007) 'Participatory research for sustainable development in Vietnam and Thailand: from a static to an evolving concept' in Franz Heidhues, Ludger Herrmann, Andreas Neef, Sybille Neidhart, Jens Pape, Pittaya Sruamsiri, Dao Chau Thu and Anne Valle Zárate (eds.) *Sustainable Land Use in Mountainous Regions of Southeast Asia: Meeting the Challenges of Ecological, Socio-economic and Cultural Diversity*, Berlin: Springer.

Rivas, Livardo and Federico Holmann (2000) 'Early adoption of *Arachis pintoi* in the humid tropics: the case of dual-purpose livestock systems in Caquetá, Colombia', *Journal of Livestock Research for Rural Development* 12(3), available at http://www.cipav.org.co/lrrd/12/3/riva123.htm (retrieved 12 October 2007).

Schuler, Ulrich, Chalathon Choocharoen, Peter Elstner, Andreas Neef, Karl Stahr, Mehdi Zarei, and Ludger Herrmann (2006) 'Soil mapping for land-use planning in a karst area of N Thailand with due consideration of local knowledge', *Journal of Plant Nutrition and Soil Science* 169: 444–52.

Smale, Melina, Mauricio R. Bellon, Jose A. Aguirre, Irma Manuel Rosas, Jorge Mendoza, Ana M. Solano, Romero Martínez, Alejandro Ramírez, and Julien Berthaud (2003) 'The economic costs and benefits of a participatory project to conserve maize landraces on farms in Oaxaca, Mexico', *Agricultural Economics* 29 (3): 265–75.

Sumberg, James, Christine Okali, and David Reece (2003) 'Agricultural research in the face of diversity, local knowledge and the participation imperative: theoretical considerations', *Agricultural Systems* 76 (2):739–53.

Talawar, Shankarappa and Robert E. Rhoades (1998) 'Scientific and local classification management of soils', *Agriculture and Human Values* 15 (1): 3–14.

Tinoco-Ordónez, Roberto (2003) 'Steps Towards Sustainable Agriculture: An Ethnopedological Soil Survey in a Limestone Area of Northern Thailand', unpublished Master's thesis, Stuttgart: University of Hohenheim.

van Veldhuizen, Laurens, Ann Waters-Bayer, Ricardo Ramirez, Debra A. Johnson, and John Thompson (eds.) (1997) *Farmers' Research in Practice: Lessons from the Field*, London: Intermediate Technology Publications.

Wünscher, Tobias, Rainer Schultze-Kraft, Michael Peters, and Livardo Rivas (2004) 'Early adoption of the tropical forage legume *Arachis pintoi* in Huetar Norte, Costa Rica', *Experimental Agriculture* 40 (2): 257–68.

Participatory research practice at the International Maize and Wheat Improvement Center (CIMMYT)

Nina Lilja and Mauricio Bellon

This study assessed the extent to which participatory methods had been used by CIMMYT, and how the scientists perceived them. Results suggest that participatory approaches at the Center were largely 'functional' – that is, aimed at improving the efficiency and relevance of research – and had in fact added value to the research efforts. The majority of projects surveyed also placed emphasis on building farmers' awareness. This is understandable if we think that the limiting factor in scientist–farmer exchange is the farmers' limited knowledge base. Thus, in situations such as marginal areas and in smallholder farming, exposure to new genotypes and best-bet management options would be a first requirement for effective interactions and implementation of participatory approaches.

Introduction

The International Maize and Wheat Improvement Center (CIMMYT) is an organisation devoted to the development of improved maize and wheat germplasm for the developing world, with a growing emphasis on addressing the needs of the poor. CIMMYT is increasingly using participatory methods in its research. However, there had not been any systematic assessment of the extent to which participatory research methods and approaches have been used, and how the scientists who rely on them perceive them (in terms of both their benefits and limitations), with a view to reflecting critically on how participatory research can make an even better contribution to CIMMYT's mission. To address some of these gaps, we looked at the use of participatory methods and approaches in the research process from the perspective of the CIMMYT scientists who adopt them.

This study[1] takes a broad look at these issues and records: (1) what is considered participatory research; (2) how it is implemented across CIMMYT projects; and (3) some of the lessons learned by scientists involved in these projects. The study has five broad research questions, formulated after a review of the relevant literature. The specific research questions are:[2]

1. What are the main characteristics of the projects using participatory approaches?
2. What types of participatory approach do the projects use?
3. What are the researchers' opinions about the usefulness of participatory methods, and what are their skills in using them?
4. Does the institutional and external environment support or constrain participatory research at CIMMYT?
5. What are the benefits and costs of participatory research?

Through a survey conducted in 2004, CIMMYT scientists reported on projects that they considered as having a participatory component; thus, the projects included in this study were self-selected. The range of the study was broad, since there was great variation in the characteristics and types of participatory research for which researchers provided information. The survey instrument allowed characterisation of the projects, but not further critical analysis of the quality or the appropriateness of the research methods applied. This was not an objective impact assessment, because linking the use of participatory research to specific impacts on farmers' livelihoods is complex and requires intermediate steps. One fundamental step is to understand and document how participatory research is perceived and used by scientists within the organisational context. This was the scope of the present study, which was intended mainly for institutional review and learning purposes.

Methodology

Externally defined criteria were not used for selecting participatory projects. Rather, in September 2004 an open call was issued for all CIMMYT staff to provide information about current and completed participatory research projects.

We used a structured survey to elicit information about the five research questions regarding project background, type of participatory approach used, its impacts, and respondents' personal views about participatory research in general. This survey form [3] was developed on the basis of a review of relevant literature (Lilja and Bellon 2008).

Information was received about 19 projects from 18 scientists.[4] Fifteen respondents were male, and three were female. Five respondents were social scientists and 13 were biophysical scientists. Sixteen of the projects involved farmer-participatory research; three projects that did not have direct *farmer*-participatory components were included because they employed participatory research with other stakeholders (national-programme scientists and seed agronomists).[5]

It is assumed that the respondents were knowledgeable about the projects, and were either actual or *de facto* leaders of the project. It is also reasonable to assume that because only currently employed CIMMYT staff were surveyed, some completed participatory projects were omitted (if the project leaders or participating scientists had left CIMMYT).

This is a qualitative study. While we provide some quantitative information on the responses received, as a general reference, these provide only a rough indication of the consensus or lack of it among respondents.

Results and discussion

Characteristics of participatory projects

The most commonly cited project goal was to increase productivity (broadly defined, but especially improved performance under various stresses). The main motivation for using participatory methods was to understand farmers' preferences better. Primary beneficiaries of these research projects were marginal farmers, but beneficiaries were not generally differentiated by sex. The 'average' participatory research project lasted for less than five years, had an annual budget of less than US$ 100,000, worked in Asia or Africa, and had six project sites, involving 400 farmers and eight scientists. That said, there was a great range and diversity among the projects.

CIMMYT participatory research projects can be viewed as collaborative activities that bring together scientific and local knowledge and all stakeholders to improve the *status quo*. The biggest obstacle to participatory research is an approach in which beneficiaries are thought of as objects of research and not as actors. Of the 19 projects surveyed, 15 were aimed at farmers, but only one specified multiple beneficiaries. Given that nearly three-quarters of the projects also stated that the motivation for stakeholder participation was to understand farmers' preferences and constraints better, this lack of recognition of multiple beneficiaries (especially the scientists) may be due to the conventional notion of 'project beneficiaries' as synonymous with 'end-users of the technology', with less emphasis placed on benefits to scientists.

It is well documented that most agricultural innovations affect men and women differently (Doss 1999). There was a noticeable absence of specific gender focus among the projects. This does not necessarily imply exclusion of gender concerns by the projects in actual research activities, only a lack of disaggregation of beneficiaries by sex. Only a few projects targeted gender differences explicitly. One other project had used a 'whole family training' approach, which included the wife and another adult female household member.

Type of participatory research approach used

The type of participatory research conducted influences the outcome of the research process. The type of participatory research is shaped by the stage at which stakeholders are involved and the types of activity in which the stakeholders are involved (Johnson *et al.* 2003; Morris and Bellon 2004). The type of participatory research used is a direct result of the motivation (or objective) to involve stakeholders in the research – that is, whether the project has primarily functional or empowering objectives (Lilja and Bellon 2008). Functional participatory research aims to improve the efficiency and impact of agricultural research and technology development: for example, the identification of traits that can guide crop breeders' work. Empowering participatory research aims to enhance local capacity to analyse problems, and seek out and develop solutions.

Functional and empowering objectives are not mutually exclusive, but (in a particular project) greater emphasis is typically placed on one rather than another. In relation to the development of agricultural technologies and information, empowering can mean giving farmers the ability to take more control of the technology options available to them and make informed decisions about their farming practices. Participatory *approaches* with either functional or empowering objectives can have both functional and empowering types of *outcome* associated with them. In economic development, the empowering approach focuses on mobilising the self-help efforts of the poor and is less often associated with the use of a single type of participatory activity or tool.

In this survey, half of the projects applied participatory tools either in priority setting only or in technology testing alone, while the other half used participatory tools in more than one stage of the research process. Most of the projects (15) used a single participatory tool. These two facts combined (stage and methods) can be used to characterise the types of participatory research used, which will affect the outcomes of the research processes. The majority of CIMMYT projects surveyed appeared to be associated with functional types of participatory method, but we do not have the necessary information to directly link the use of methods to types of outcome.

For three-quarters of the projects, the primary reason for involving stakeholders was to increase the relevance of research and to make research and extension more demand-driven by better understanding of farmer preferences and constraints, and to use farmers' knowledge in technology evaluation and development. This can be interpreted as a functional approach with emphasis on co-learning. For the other quarter of the projects, the main motivation was to involve stakeholders in technology dissemination and to improve awareness and hence the reach of technology. Our interpretation is that these projects also have a functional, but more action-oriented approach, where emphasis is placed on translating new knowledge into improved farmer practice through participatory dissemination. Both of these functionally motivated approaches may also lead to greater farmer empowerment.

Quality of science in participatory research

In understanding the potential advantages of participatory approaches, there are also methodological issues in blending scientific and local knowledge that need to be considered (Campbell 2001; Berardi 2002). Rather surprisingly, none of the scientists in the survey said that participatory research would be best suited for all aspects of the research continuum; about two-thirds said it was best suited for technology evaluation, testing, and dissemination; one-third said that participatory approaches were best suited for priority setting. The answers may reflect two opposing attitudes and situations: one in which research has identified 'suitable technology options', and interaction with farmers is believed to increase adoption because farmers need to learn about the options through experimental learning combined with better farmer-to-farmer dissemination. The other situation may reflect the opinion that farmers have a key role in defining the research priorities, but fewer roles in developing the technology options.

Most participatory research at CIMMYT has a functional objective, aimed either at increasing the efficiency of the research process to generate 'better' research products, or at fostering the diffusion of these products by enhancing the awareness and knowledge of potential beneficiaries about them. For example, as the physical and economic resource bases of different groups necessitate tailored research, functional approaches allow scientists to direct their research according to the needs of the specific groups of farmers and specific environments. Farmers can assure scientists that they are assessing trade-offs among variety traits and management practices 'correctly and under real-life conditions', which ensures greater adoption of innovations by the farmers.

More empowering objectives to participatory research would aim to increase farmer knowledge and skills, so that they can participate more fully in the collaborative breeding efforts and improve their own personal efforts. Empowering approaches to participatory research are not merely about increasing farmers' awareness. As most CIMMYT projects were concerned with understanding farmer preferences, there was less focus on targeting, equity concerns, or building participants' skills. For example, many scientists felt at the onset of the project that farmers needed to learn about new varieties and management practices. The apparent emphasis on building farmers' awareness is understandable if we think that the limiting factor in scientist–farmer exchange is the farmers' (limited) knowledge base. Thus, in situations such as marginal areas and in smallholder farming, exposure to new genotypes and best-bet management options would be a first requirement for effective interactions.

The fact that the majority of the respondents said that farmers needed more information may be viewed in two different contexts. On one hand, it may reflect the prior understanding of the farmers' specific needs and constraints for improved varieties, and management and resource-conservation techniques. On the other hand, it may reflect some scientists' biases about how the formal-sector research has already fully identified solutions to the specific farmer problems and constraints: four-fifths of the respondents said that by the start of the projects it was already determined that farmers needed to learn more information.

Participatory research has its origins in qualitative methods, and the use of these methods is most often associated with social scientists. However, the majority (13) of the survey respondents at CIMMYT were biophysical scientists. The survey method did not allow assessment of scientists' competence in participatory methods, as doing so would have required more detailed individual interviews and field observations. Instead, we asked about their 'comfort level' in using the participatory methods, but this should not be understood as a proxy for competence in their use. There was a very high confidence level in the use of participatory methods, although hardly any of the respondents had any training in participatory research. Some of the answers reflected the common attitude that the use of participatory methods is 'common sense', requires little or no formal training, and is easy for 'people-oriented' researchers.

There seemed to be a positive perception of participatory research among the majority of practitioners, who considered participatory methods most appropriate for technology and varietal evaluation and testing. Most scientists were self-taught in participatory research methodology (they did not seem to have any formal training in the methods and approaches of participatory research), felt comfortable using the methods after one year, and with extending the methods to others after two years. Rather surprisingly, despite this apparent comfort in extending the methods, combined with a perception that colleagues at CIMMYT appreciated participatory research, the majority of scientists said that they had never been asked to advise on participatory research. This suggests that there was a lack of communication and sharing of knowledge and experience. This may be problematic, since, by not communicating, they may have been 'reinventing the wheel', or their work may not have been as efficient as it could have been. This suggested a lack of institutional space to share and learn from the extensive and valuable experience being generated by CIMMYT scientists in this respect.

Three facets of CIMMYT participatory research were expected to foster and promote peer acceptance of new approaches and allow for faster scaling up in research efforts: biophysical scientists (not only social scientists) were involved in participatory projects; there seemed to be an interdisciplinary approach in most projects; and these projects seemed rather well connected to the pre-existing network of scientists and other projects.

Institutional issues

For a public research institution with a global mandate, such as CIMMYT, the functional aim of participatory research is more relevant, given a mandate to produce research products that are globally applicable.[6] But this is a challenge, since the value of participatory research lies in its focus on local issues and contexts, which can result in research products of very limited applicability. This is also a limitation to a widespread use of the empowerment aim of participatory research, since it could only be done among a few communities or places, which clearly is not cost-effective. It is therefore not surprising that there is an emphasis on evaluating, adapting, and extending technologies developed previously by the formal research system among the projects analysed here. A model of participation in which farmers are actively involved in research is often set as an 'ideal type'. The evidence from this study suggests that while information flows both ways between scientists and farmers, the dominant information flow is still top–down or researcher-directed; there is no clear, broad trend towards client participation in the testing stages of the research process. This is consistent with studies in Nepal (Gauchan *et al.* 2000; Biggs and Smith 2003). This could be seen as participatory research (with its two-way information flows) conducted within a linear, pipeline model of innovation, but it also provides evidence of the challenges of balancing local contexts with the need for broader applicability. Nevertheless, the evidence provided here still suggests the use of participatory research within the context of a dominant supply-driven agenda.

The survey results show limited interaction among CIMMYT participatory projects. One possible explanation for this is that there is sometimes a tendency for individual scientists or projects to 'trademark' their participatory methodology with an excessive focus on participatory acronyms (Berardi 2002). This is good to the extent that it shows a sense of ownership of the participatory methodologies developed, but it can be problematic if it leads to seeing the development of technical solutions as a separate, isolated research effort. However, an alternative explanation is that there is a lack of an institutional structure to learn, share, and reflect systematically on the lessons generated by the application of participatory methods. This is probably the most immediate institutional problem faced by CIMMYT if it is to realise the benefits from participatory research more fully.

Benefits and costs

Scientists provided their perceptions of the differences that participation made to outcomes[7] and the expected outcomes had participation not been used. At least conceptually, these perceptions provided a sort of counterfactual regarding participation. Box 1 presents a synthesis of the outcomes derived from these perceptions. These clearly are not impacts, since the links to changes in the beneficiaries' livelihoods have not been documented or measured; however, they are fundamental, being a necessary but insufficient condition for impact.

Box 1: Outcomes associated with participatory research at CIMMYT

- Increased diversity.
- Demonstrated the value of diverse maize landraces to farmers.
- Demonstrated the farmers' preference for open-pollinated varieties over hybrids, particularly under stress conditions.
- Provided farmers with access to seed and promoted faster adoption.
- Made farmers aware of new varieties and fostered faster adoption.
- Provided farmers with varieties with valued traits.
- Increased the ability of farmers to evaluate resource-conserving technologies and assess their benefits.
- Minimised the error of developing varieties that farmers do not want (or with traits they do not value) or are not relevant for their preferences and circumstances.
- Developed research products (varieties) that are relevant for users who value multiple characteristics.
- Understood the constraints faced by farmers; established baselines to assess impacts.
- Made the research process more efficient by identifying pathways to reach farmers.
- Understood the context in which new technology has to operate.
- Allocated technologies to appropriate niches in the farming system.
- Provided farmers with information from other stakeholders that have impact on their lives.

The identified outcomes could also become the subject of more rigorous study and of monitoring, but this would be a next step. Furthermore, since the outcomes have been identified, it may be easier to make predictions about the potential impacts that may be associated with them. These predictions could then be the basis for more rigorous impact analyses which link research process and outputs to livelihood changes. This could be the basis for a more in-depth quantitative study on the impact of participatory research at CIMMYT. Most importantly, such a study would also have to address the perceptions of the beneficiaries and other stakeholders of the outcome of their participation.

The benefits of a research project are evaluated in comparison with the costs of the research. The survey results show that there was a diversity of views about the costs among practitioners: some considered that there were additional costs, while others did not think so. Furthermore, it is clear that in many cases comparing the costs of participatory research with those of more conventional research is difficult, because the approaches may be so different that it is not really meaningful to compare them. In any case, it seems that from the perspectives of CIMMYT practitioners, participatory research may not entail additional costs or, if it does, the results justify the expense.

It is difficult to compare the costs of participatory research with those of 'conventional' research, because in reality a research process often includes both conventional and participatory activities. It would be erroneous simply to conclude that participatory research is more costly than conventional research. In reality, the share of the overhead and personnel costs often remains fixed, and operations are adjusted according to the availability of funds. Participatory research usually affects the operational costs the most – and not always by increasing them, especially if it replaces some other activities. If participatory research is implemented as an add-on activity, then the research costs are likely to increase (Lilja and Aw-Hassan 2003).

Half of the survey responses on the 'impact' of participatory research provided examples of 'impact' of variety and technology evaluation (14 of the 27 'impacts' reported) – improvement over the *status quo* in understanding farmers' preferences, experiences, needs, social and production constraints, as well as solutions that they may offer to the collaborative research process. The results imply success in shortening the time-lag between technology development and its adoption, which has important implications for overall returns to research investment.

Examples of 'impact' of surveys (11 of the 27) (elicitation of farmer preferences and knowledge) and diagnostic needs-assessment show the benefits of broader socio-economic information, and how it can help to determine who the actual beneficiaries will be in the various social strata or resource-dependent groups, and what the specific preferences and constraints are for each. Such information can also help to reveal, in advance, the unintended potentially negative and positive impacts of a project on different groups within the project area.

Concluding remarks

The financial resources associated with what is claimed to be participatory research at CIMMYT are rather surprising: approximately US$ 9 million per year. While this sum refers to the research that has participatory components and may not reflect the specific resources invested in participatory activities *per se*, this level of investment clearly indicates that participatory research is more than a marginal activity in the institute. CIMMYT may need to consider investing additional resources to create a more conducive environment for its scientists to share their experiences and learn from each other, and in doing so add value to this research endeavour, or else participatory research may become a meaningless, catch-all term used for data collection or the analytical phase of research. Furthermore, this may also require more investment in documenting the outcomes and impacts of participatory research at CIMMYT. We believe that, by identifying the projects and the outcomes associated with participation, the research reported here is laying the groundwork for further advances in this area.

Acknowledgements

We gratefully acknowledge the valuable contributions of the following scientists who replied to the survey: Marianne Banziger, David Bedoshvili, David Bergvinson, Hugo De Groote, Dennis Friesen, Luz M. George, Chandra Gurung, Scott Justice, Craig Meisner, Alexei Morgunov, Julie Nicol, Guillermo ('Memo') Ortiz-Ferrara, Kamal Raj Paudyal, Zondai Shamudzarira, Carlos Urrea, Steve Waddington, and Patrick Wall. We also appreciate useful suggestions from Barun Gurung, Kevin Pixley, and Robert Tripp, and editorial assistance and comments from Mike Listman and Guy Manners. Our sincere thanks to CIMMYT for allowing us to use material from Lilja and Bellon (2006), and to USAID for generously supporting the study through linkage funding.

Notes

1. This article is based on Lilja and Bellon (2006).
2. Because these results came from a survey of project scientists, we were not able to analyse local people's willingness and ability to participate, which are hypothesised to affect the type of participatory research that is conducted. Similarly, we do not have information on local people's opinions and views about the other research questions: the local knowledge represented, the external factors, and benefits and costs of the project.
3. The survey form with detailed questions is available from the authors.
4. In all projects except one, only one person filled out the survey. In that one project, two people filled out the survey form jointly. Also, one person provided information about three projects.
5. Further details about the projects are provided in Lilja and Bellon (2006).
6. For a further discussion of the advantages and disadvantages of functional and empowerment aims of participatory research in the context of public agricultural research institutions, see Hellin *et al.* (2008).
7. 'Outcomes' are the changes resulting from uses of outputs by stakeholders and clients (for example, changes in knowledge, attitudes, policies, research capacities, and agricultural practices), whereas 'impacts' are the longer-range social, environmental, and economic benefits that are consistent with the institution's mission and objectives (for example, increased agricultural productivity, improved food distribution).

References

Berardi, G. (2002) 'Commentary on the challenge to change: participatory research and professional realities', *Society and Natural Resources* 15 (9): 847–52.

Biggs, S. and S. Smith (2003) 'Paradox of learning in project cycle management and the role of organizational culture', *World Development* 31 (10): 1743–57.

Campbell, J. R. (2001) 'Participatory Rural Appraisal as qualitative research: distinguishing methodology issues from participatory claims', *Human Organization* 60 (4): 380–89.

Doss, C. R. (1999) 'Twenty-five years of research on women farmers in Africa: lessons and implications for agricultural research institutions; with an annotated bibliography', *CIMMYT Economics Program Paper* 99-02, Mexico, DF: CIMMYT.

Gauchan, D., M. Joshi, and S. Biggs (2000) 'A new NARC strategy for participatory technology development and linkages with multiple actors', in M. Joshi, D. Gauchan, and N. Thakur (eds.) *Proceedings of the 5th National Outreach Research Workshop, 30–31 May 2000, Outreach Research Division, Nepal Agricultural Research Council (NARC), Kathmandu, Nepal.*

Hellin, J., M. R. Bellon, L. Badstue, J. Dixon, and R. La Rovere (2008) 'Increasing the impacts of participatory research', *Experimental Agriculture* 44 (1): 81–95.

Johnson, N. L., N. Lilja, and J. A. Ashby (2003) 'Measuring the impact of user participation in agricultural and natural resource management research', *Agricultural Systems* 78 (2): 287–306.

Lilja, N. and A. Aw-Hassan (2003) 'Benefits and Costs of Participatory Barley Breeding', background paper for poster presented at the International Agricultural Economics Association meeting in Durban, South Africa, 16–22 August.

Lilja, Nina and Mauricio Bellon (2006) 'Analysis of Participatory Research Projects in the International Maize and Wheat Improvement Center', Mexico, DF: CIMMYT.

Lilja, Nina and Mauricio Bellon (2008) 'Some common questions about participatory research', *Development in Practice* 18 (4&5): 477–486.

Morris, M. L. and M. R. Bellon (2004) 'Participatory plant breeding research: opportunities and challenges for the international crop improvement system', *Euphytica* 136: 21–35.

Making poverty mapping and monitoring participatory

Li Xiaoyun and Joe Remenyi

The real experts on poverty are poor people, yet the incidence and trends in poverty are usually measured by the use of official economic indicators assumed by researchers to be relevant. Poor householders themselves distinguish between subsistence and cash income. In a 'self-assessed poverty' exercise, poor villagers in rural China specified and weighted key poverty indicators. Eight key indicators describing three basic types of poverty were isolated and used to construct a participatory poverty index (PPI), the components of which provide insights into core causes of poverty. Moreover, the PPI allows direct comparison of the incidence of poverty between villages – differences in social, cultural, and environmental characteristics of each village notwithstanding. As a result, the PPI offers an objective method of conducting poverty monitoring independently of physical and social features. This article provides a brief description of the PPI and the data needed to construct a village-specific PPI.

Introduction

In its report *Poverty Trends and Voices of the Poor*, the World Bank (2001a: 3) defines poverty as 'a multidimensional phenomenon, encompassing inability to satisfy basic needs, lack of control over resources, lack of education and skills, poor health, malnutrition, lack of shelter, poor access to water and sanitation, vulnerability to shocks, violence and crime, lack of political freedom and voice'. This article examines the gap between 'the way development agencies measure poverty' and the reality of 'how poor people experience and understand their poverty'.

Early in 2001, the Beijing office of the Asian Development Bank (ADB), in partnership with China State Council's Leading Group Office for Poverty (LGOP), set about identifying a means by which the reality of poverty at the grassroots, especially chronic hard-core poverty in village China, could be measured, monitored, and addressed. China's experimentation with Village Poverty Reduction Planning (VPRP)[1] is the outcome, establishing a widely accepted theoretical basis for participatory poverty-reduction planning – including poverty mapping, poverty-reduction priority setting, and participatory monitoring of changes in the incidence of poverty – across much of China. This article reports on the replicable framework of field-tested indicators[2] underlying VPRP.

Background trends in the incidence of poverty in rural China

The decline in poverty in China since 1980 is one of the great achievements in modern development. In 1978 (the year that China began to open its economy to the world), the head-count index showed at least 250 million poor in rural China. Thirty years later, this figure is down to less than 30 million – almost a ten-fold decline.[3]

Substantial gains in poverty reduction notwithstanding, official data on poverty trends in China since the mid-1990s showed that progress in reducing village poverty was not keeping pace with growth in the rest of the economy. Equity problems are emerging, to the detriment of pro-poor growth. The data indicate that the poorest households in rural China are being increasingly marginalised, and progress in reducing poverty has stalled, possibly even reversed. In 1985–2000, rural off-farm employment grew by less than five per cent, while the numbers employed in agriculture fell by ten per cent, increasing the challenge of labour mobility as a key livelihood strategy for poor households.[4]

The research to test VPRP's participatory alternatives to pro-poor planning in rural China has added insights into how participatory strategies in poverty-reduction planning can redress the slide of village China into chronic hard-core poverty. Field tests of VPRP show that the top–down nature of extant poverty mapping and poverty-reduction planning processes has contributed significantly to both the neglect of interventions that are directly relevant to poverty alleviation at village level, and the leakage of resources to interventions that have not benefited the rural poor. VPRP presents an opportunity to significantly improve poverty targeting by addressing symptoms of poverty that are directly relevant to the reasons why village poverty persists, especially in the poorest households.

The VPRP approach to poverty measurement and monitoring

China's chronically poor are the true experts on what it means to be poor, the constraints that prevent them from escaping from their poverty, the opportunities that they would like to take up as investments in their own betterment, and the key areas of assistance that they need. VPRP asked poor rural villagers in Hebei Province to select the poverty indicators that they felt best described their poverty. A common subset of eight key indicators covering three critical types of poverty emerged. The combination of these eight indicators represents a view of *how poverty is experienced by poor households in village China*, as assessed by respondents. This is not to say that villagers consulted did not identify other indicators – they did. But it is these three types of poverty and the eight indicators of these three types of poverty that have been constants in responses received. It is this subset that forms the basis of VPRP and the Participatory Poverty Index (PPI).

When the first trials of the VPRP framework were undertaken in Fengning County, Hebei Province, a series of 11 group meetings was held. Poor villagers and officials were asked for their views on how best to measure and monitor poverty, and how best to address the problems that give rise to their poverty. In each village, meetings were held with at least 20 villagers. At times, the meetings resembled a social gathering as individuals shared anecdotes and animatedly discussed and argued the point they were trying to make. A single meeting could easily extend through the whole morning or afternoon. Typically, each village meeting was held in the village primary school or health centre. In addition, two township meetings were held, involving almost 200 county and township representatives from the 26 villages reputed to be the poorest among Fengning County's 309 villages. The results of these consultations are summarised in Table 1.

The rank order of poverty indicators shown in Table 1 was arrived at after a lengthy consultation with respondent groups, who considered a range of possible indicators. It is significant

Table 1: Poverty indicators selected by 180 poor villagers and 252 village and county officials, Fengning County, Hebei Province, China

Poverty indicator in rank order		Priority accorded by:		
Officials villages		**Fengning County officials (122)**	**Township (78) and village (52) officials**	**Poor villagers (180) from 9 villages**
1.	Cash receipts per person per year	2	1	1
2.	Grain production per person per year	1	2	2
3.	Access to an all-weather access road	5	3	3
4.	Easy access to good-quality drinking water	4	6	4
5.	House quality (roof and exterior walls)	3	5	7
6.	Access to reliable electricity supply	6	4	8
7.	Days lost due to women's ill health	7	7	5
8.	Children's access to education	8	8	6
9.	Average arable land per person	4	10	9
10.	Frequency of natural disasters	0	9	11
11.	Availability of irrigation	9	12	0
12.	Degraded local ecology	0	0	10
13.	Access to poverty loans	0	11	0

that respondents chose to ignore many indicators, concentrating instead on the overlap in the top eight indicators selected by each group. This outcome has been crucial for the design and further development of VPRP as a practical and meaningful approach to the use of participatory strategies in the design, implementation, and impact monitoring of village poverty-reduction policies in China.

The very much higher priority given by poor villagers and village officials to the existence of an all-weather access road is indicative of the influence that such an asset has on those who suffer most when access to nearby markets, emergency assistance, off-farm job opportunities, etc., is restricted by transport constraints. In like manner, the higher priority given by poor villagers to the health of the able-bodied women in their households can be directly linked to the critical role that poor villagers recognise is played by women in household livelihood, asset creation, and resource-management activities. However, it is the similarities that are most outstanding. In contrast to the typical official approach to poverty monitoring and mapping according to the level of subsistence production or value of per capita income attributable to each household member, poor villagers and officials who work most closely with poor villagers distinguish between the cash and non-cash components of income as a measure of poverty. Even among the most deeply subsistence-based households, the chains of poverty are rarely loosened if there is no added access to cash-flow – trends in subsistence production notwithstanding. Consequently, when seeking to monitor the impact of public-sector assistance to poor households, it is important to deliberately seek out ways in which cash flow into poor households can be augmented.

The first eight poverty indicators in Table 1 describe three key types of poverty: livelihood, infrastructure, and human-resource poverty. It is this subset of indicators that the VPRP field trials focused upon as the core data needed for effective participatory poverty-reduction planning at village level. Table 2 presents these eight indicators according to the form of the data that were collected in field trials.

These eight indicators give a unique insight into poverty as it is experienced and understood by poor villagers. The immediate livelihood components draw attention to the importance of grain production as a major source of nutritional security, but the higher priority given to financial liquidity underscores the fact that money matters when you are poor, further highlighting the importance that the poor give to increases in the value of surplus marketable production, total cash earnings from off-farm employment, and financial safety-net payments.

Causes of poverty traps can be legion, but the reality is that many constraints are directly linked to the level of investment that has been made in the accumulation of local infrastructure assets with public-goods qualities. For example, an all-weather access road is a conduit of information for everyone about markets and potential employers. Similarly, public-sector investment in potable water can not only boost household labour supply but also slash health costs linked to poor-quality drinking water. Access to a reliable supply of electricity, however, is an indicator of a totally different order. This indicator has little to do with the consumption of a public good. Rather, it reflects the extent to which poor villagers are able to plan and realise their entrepreneurial production activities.

A Participatory Poverty Index (PPI)

VPRP field trials not only consulted poor villagers on how best to map poverty, but also engaged them in the process of quantifying each indicator and attaching relative weights to each indicator, including a weight for each of the three key *types* of poverty. As a result, the VPRP process allowed for the construction of a composite Participatory Poverty Index (PPI) which allows direct comparisons of the depth of poverty across villages or larger communities. Uniquely, because the PPI is participatory, it allows for poverty mappings that reflect poverty from the perspective of the poor themselves, as contrasted with poverty-assessment methods that are imposed from the point of view of the researcher or the policy maker.[5]

Table 2: Eight key indicators for three core types of poverty

Poverty indicators in rank order	Measure
1. *Livelihood Poverty*	
a. Cash receipts per person per year	Yuan
b. Grain production per person per year	kg
c. House quality (roof and exterior walls)	% brick
2. *Infrastructure Poverty*	
a. Access to an all-weather access road	Days without access
b. Access to reliable electricity supply	Days with interrupted supply
c. Easy access to good-quality drinking water	Hours spent collecting water
3. *Human Resource Poverty*	
a. Women's morbidity	Days lost to illness, females >12 years
b. Children's education	% of eligible children in school

The data associated with each of the eight poverty indicators provide a reality-based picture of what needs to change if the incidence of poverty is to fall and people in village China are to genuinely *feel* less poor.[6] Consider the following examples.

Among the three indicators of livelihood poverty, first importance is given to household cash flow. The policy implication is clear: poor people want policy makers and poverty-reduction planners to give greater priority to increasing their access to cash receipts or cash-earning opportunities. Cash for work is unequivocally preferred over food for work. Similarly, the poor recognise that food insecurity, the second livelihood indicator, is not solely about the inability of poor households to produce enough food for their own household use, but also about the absence of cash resources to be able to *buy in* the shortfall. That the third livelihood indicator should focus on the quality of housing is not surprising: the quality of housing is an excellent proxy for household poverty. Lack of access to potable water is important for sustainable poverty reduction. Time spent fetching water is 'dead time', restricting the available labour power for household or farm production. Illness of household members from water-borne diseases is a drain on household savings and household cash reserves, and increases the household dependency rate. Being cash-poor, food-insecure, and unwell compounds the difficulty of escaping the mire of poverty unassisted.

The PPI incorporates the views of poor householders on what they believe are the critical causes and impacts of *their* poverty on the quality of *their* lives.[7] Poor villagers are able to rank the eight poverty indicators according to which are most and least relevant to the ways in which they experience poverty. These rankings, which are not imposed by the researchers, but reflect the revealed priorities of poor villagers, are recorded using a scale of 1 (most important) to 5 (least important), which poor villagers are asked to attach as weights to each indicator and each type of poverty. On the basis of this range, each indicator can be converted to a common base, from a low of five to a maximum of one.[8]

Relative poverty is influenced by local knowledge of factors such as the importance of subsistence production to local survival strategies; the relevance of population density in the village's unique geographic context; and the role that distance of village households from townships, for example, plays in determining local income-generation opportunities. Poor villagers are well aware of these influences on the quality of their lives. As a result, it is poor villagers who are best able to determine the poverty range that should apply to each indicator. The poverty range is independent of the official income-based poverty line, which sets a monetary (¥) limit to per capita income. The poverty range may overlap with the official poverty line, but it can also be either above or below the official estimate.

One might speculate that the participatory nature of the VPRP process might undermine the veracity of the poverty-mapping results, because the process may present villagers with the opportunity to exaggerate their poverty so as to maximise the chances that their village will be selected for public-sector assistance. In an effort to avoid this possibility, VPRP includes a peer-review step, which is designed to allow village representatives to review and comment on the baseline data collected from all villages in their area. During these peer-review sessions, villagers and village officials are able to challenge the veracity of individual indicators, relative to what is shown for their own village. Eventually, a consensus does emerge about the relative position of one village compared with another for each of the eight indicators.

VPRP field trials showed that villagers are pragmatic and realistic; not a single example of exaggeration was encountered in the many meetings held across Hebei Province. Rather, poor villagers took the opportunity to participate in VPRP as occasions to draw attention to their needs and the key problems that must be overcome for poverty in their village to be reduced, if not abolished. The participatory process appears to insert an element of 'robustness' into poverty-reduction planning, in particular to the PPI-based poverty mapping.

Calculating the PPI

The arithmetic needed to calculate a PPI for each village is not complex. This is important, because it is village, county, and township officials especially for whom the PPI ought to be an important new tool in the fight to abolish hard-core village poverty. The weights provided by villagers are adjusted to a base of 100 for easy comparison of the incidence of poverty across villages. The closer the PPI is to 100, the more severe and deeper the incidence of poverty.

The formula for the PPI can be expressed algebraically as follows:

$$PPI = \left[\sum (I_i^* w_{it})^* W_t \right]^* 20$$

where:

$W_{t=1-3}$ = poverty weights for each component of: Livelihood (W_1), Infrastructure (W_2) and Human Resource (W_3) poverty. $\sum W_{t=1-3}$ = 1. $I_{i=1-8}$ = eight poverty indicators converted to within a poverty range of 5–1. w_{it} = poverty weights for each poverty indicator relevant to each type of poverty where t = 1–3 and i = 1–8 and $\sum w_{it}$ = 1.

In expanded form, the PPI is written as:

$$PPI = [W_1(I_1w_{11} + I_2w_{12} + I_3w_{13}) + W_2(I_4w_{21} + I_5w_{22} + I_6w_{23}) + W_3(I_7w_{31} + I_8w_{32})]20$$

Table 3: Calculating the Participatory Poverty Index for three villages in Fengning County, Hebei Province

Poverty I_i	Poverty Range			Poverty Type Weights W_t			Poverty Indicator Indicator Weights w_{it}		
	Village								
	1	*2*	*3*	*1*	*2*	*3*	*1*	*2*	*3*
Livelihood Poverty				0.35	0.38	0.34			
Grain production (I_1)	4	5	5				0.37	0.41	0.35
Cash receipts (I_2)	4	3	4				0.39	0.38	0.40
Quality of housing (I_3)	3	5	5				0.24	0.21	0.25
Infrastructure Poverty				0.33	0.31	0.34			
Access to potable water (I_4)	4.5	2	1				0.41	0.41	0.40
Access to reliable electricity (I_5)	2.5	1	1				0.24	0.28	0.20
Access to all-weather road (I_6)	3.5	5	5				0.35	0.31	0.40
Human Resource Poverty				0.32	0.31	0.32			
Women's health	5	4	3				0.50	0.47	0.53
School attendance	3	5	2				0.50	0.53	0.47
Participatory Poverty Index (PPI)				**77**	**76**	**64**			

Table 4: Ranked PPI estimates for 309 villages in Fengning County, Hebei Province

Town	Village	PPI	Town	Village	PPI	Town	Village	PPI	Town	Village	PPI	Town	Village	PPI	Town	Village	PPI
Nanguan	6	86.2		6	67.4		6	61.1			58.3	Sichakou	2	64.9		13	48.5
Tucheng	1	92.6		7	73.1		7	54.3	Datan	1	67.9		3	63.1		14	48.5
	2	92.6		8	92.1		8	50.6		2	67.8		4	61.3		15	47.8
	3	92.6	Beitou	1	92.1	Caoyuan	1	62.9		3	66.1		5	61.3		16	31.0
	4	92.6		2	78.8		2	63.5		4	65.8		6	61.3	Yuershan		50.2
	5	92.6		3	78.7		3	63.5		5	65.8		7	59.1		1	57.7
	6	90.8		4	77.4		4	63.5		6	65.8		8	57.6		2	53.3
	7	89.1		5	75.9		5	60.9		7	63.2		9	56.3		3	52.9
	8	89.0		6	75.4		6	62.3		8	63.1		10	52.6		4	48.6
	9	87.2		7	74.1		7	73.4		9	63.1		11	50.1		5	48.6
	10	85.0		8	73.0		8	71.5		10	61.3		12	50.1		6	48.6
	11	81.3		9	70.2		9	63.9		11	61.0		13	46.3		7	48.6
	12	78.0		10	66.7		10	61.4		12	61.0		14	44.5		8	46.8
	13	77.6		11	49.8		11	59.6		13	57.8		1	56.7		9	46.8
	14	65.2		12	46.0		12	54.3		14	57.6	Tianqiao	1	62.7	Huangqi		49.2
Kulongsan	1	79.0		13	71.6		13	51.8		15	56.3		2	60.2		1	62.4
	2	92.0	Wudan	1	80.1		1	62.2		16	56.2		3	60.2		2	62.4
	3	92.0		2	74.8		2	84.5		17	54.0		4	58.0		3	62.4
	4	85.5		3	74.5		3	83.0		18	50.1		5	55.4		4	62.3
	5	85.5		4	72.4		4	76.1		19	50.0		6	53.2		5	59.7
	6	85.5		5	69.7		5	73.1		20	44.9		7	52.7		6	57.5
	7	85.5		6	69.5		6	63.2		21	44.9		8	50.9		7	55.3
	8	85.5		7	60.5		7	45.6		22	38.6		1	53.6		8	55.3
			Tanghe	8	71.0		8	39.1	Xiguan	1	58.1		1	67.7		9	54.9

(Table continued)

Table 4: Continued

Town	Village	PPI	Town	Village	PPI	Town	Village	PPI	Town	Village	PPI	Town	Village	PPI	Town	Village	PPI
	9	85.5		1	83.3		8	32.8		1	73.5		2	65.1		10	52.2
	10	85.5	Waigou	2	78.1			59.9		2	62.0		3	65.1		11	47.8
	11	85.5		3	76.4		1	64.7		3	60.2		4	52.7		12	44.0
	12	85.5		4	73.7		2	61.7		4	57.9		5	52.7		13	44.0
	13	85.5		5	73.4		3	60.2		5	57.0		6	51.6		14	44.0
	14	85.5		6	73.3		4	58.9		6	55.3		7	46.8		15	40.3
	15	85.5		7	72.9		5	57.5		7	55.3		8	40.3		16	30.1
	16	82.9		8	71.1		6	56.2		8	55.3		9	40.3		17	25.6
	17	82.9	Sujiadian	9	70.8	Heishan		59.1	Xueyin	9	53.4	Dage		53.6		18	25.6
	18	82.9		10	70.7		1	68.4		10	50.9		1	62.9			41.7
	19	79.1		11	66.0		2	60.2			57.3		2	62.7		1	70.7
	20	79.1		12	63.3		3	58.3		1	70.6		3	60.9		2	63.7
	21	78.9		13	62.8		4	57.6		2	66.6		4	60.9		3	60.1
	22	76.7		14	57.7		5	55.7		3	65.9		5	59.7		4	55.6
Boluonuo	23	76.7			68.8		6	54.6		4	64.1		6	53.3		5	48.6
	24	76.7	Wanyong	1	94.0			59.1		5	64.1		7	50.6		6	45.8

(Table continued)

Table 4: Continued

Town	Village	PPI	Town	Village	PPI	Town	Village	PPI	Town	Village	PPI	Town	Village	PPI	Town	Village	PPI
Xiaobazi	25	76.2		2	94.0		1	66.1		6	62.9		8	43.1		7	44.7
	26	76.2		3	91.2		2	59.2		7	59.6		9	42.9		8	42.1
	27	76.2		4	79.9		3	58.7		8	59.1		10	38.6		9	41.8
	28	76.2		5	74.2		4	52.4		9	56.9	Humayin		53.3		10	39.3
	29	72.7		6	70.6	Sirengou		58.5		10	56.6		1	72.3		11	39.3
	30	60.3		7	61.6		1	79.3		11	54.5		2	68.7		12	35.6
	31	60.3		8	53.4		2	65.2		12	53.5		3	60.2		13	35.6
	32	60.3		9	48.0		3	62.4		13	52.9		4	60.2		14	34.9
	33	60.3		10	46.0		4	59.1		14	52.9		5	57.5		15	34.9
	34	54.7		11	43.6		5	58.2		15	52.3		6	53.1		16	34.9
Wangyin		76.2	Yangmu		65.2		6	56.4		16	50.5		7	53.1		17	34.9
	1	80.1		1	79.3		7	55.5		17	50.5		8	52.7		18	34.9
	2	79.6		2	76.7		8	52.9		18	47.5		9	50.5		19	29.2
	3	78.4		3	74.5		9	52.9		19	47.5		10	50.3		20	27.3
	4	76.8		4	64.3		10	51.3			56.7		11	50.3		21	22.7
	5	74.9		5	61.1		11	50.2		1	65.7		12	48.6			

Table 5: Comparison of wealth ranking and PPI for nine villages, Fengning County, Hebei Province

Village	Wealth ranking 1 (wealthiest) to 9 (poorest)	PPI (% incidence of poverty)
Tucheng	1	50
Dongshang	2	46
Qianfoshi	3	67
Sijianfang	4	70
Sanjianfang	5	75
Liutiaogou	6	74
Zhangying	7	79
Mawopu	8	92
Liquan	9	92

The calculation of the PPI is essentially a simple arithmetic exercise, as shown in Table 3, where data for three villages from Fengning County, Hebei Province, are presented. The data shown in Table 3 were gathered as part of the field trials of County Poverty Alleviation Planning (CPAP).

Table 3 contrasts three villages with PPIs ranging from 64 to 77. These numbers can be interpreted as showing that poverty is significant in all three villages, but least severe in village 3 and most severe in village 1. Moreover, the components of the Index tell us that all three villages have a significant food-security problem, although women's health and potable water are priorities in villages 2 and 3.

The VPRP field trials have been extensive enough to allow for the calculation of PPI estimates for all 309 villages in Fengning County, Hebei Province. If one takes a PPI of 50 as a cut-off for villages in urgent need of poverty-reduction assistance, the results confirm Fengning as a genuinely poor county. More than 80 per cent of villages in Fengning County have a PPI greater than 50 (Table 4)!

PPI and wealth rankings compared

If the PPI is to stand the test of validity, it ought to bear a close relation with village wealth rankings, whereby the poorest villages are also the least wealthy villages. In order to test this relationship, wealth rankings were also conducted in nine of the villages (Table 5). A close correspondence was found. The wealth rankings, however, unlike the PPIs, are not pregnant with information and policy guidelines on why these levels of poverty exist and how the poor in villages can best be assisted to improve the quality of their lives.

The parallel between the wealth rankings and the corresponding PPIs is striking (Table 5).

Conclusion

Villages in Fengning County have PPIs that range between a high of 92.6 for the poorest villages and a low of 22.7 for the least poor. These differences can be interpreted as an indicator not only that one village is relatively poorer than another, but also of the 'depth' of the incidence of poverty in each village. It is legitimate to say that a PPI of 90 indicates a depth of poverty that is 70 percentage points more intense than a PPI of 20. In this sense, the PPI goes beyond simple head-counting. It incorporates allowances for how poverty manifests itself at the village level and in poor households. Its component parts provide planners with insights into the priority problems as perceived by villagers. This is important new information, responses to which can be expected to mobilise community support for Village Poverty Reduction Planning.

Notes

1. In the reports prepared for the ADB, VPRP was called CPAP (County Poverty Alleviation Planning) (see Li *et al.* 2002).
2. These indicators are discussed in Li *et al.* (2003) and Remenyi and Li (2004).
3. Poverty trends in China are reviewed in: UN (2006); China State Council (2001); Fan (2002); LGOP (2000, 2002); LGOP *et al.* (2000); and World Bank (2001b).
4. See especially Ravallion and Chen (2007).
5. The participatory character of the VPRP strategy and the PPI is in marked contrast to the typical poverty-assessment strategy used by development-assistance agencies – see, for example, the results of the USAID-sponsored Developing Poverty Assessment Tools Project (USAID 2004); the IFPRI–CGAP approach in Henry *et al.* (2003); and Remenyi and Kelly (2007).
6. It should be expected that different communities of poor will identify different indicators as critical to them. What is common is the insight that the selected indicators provide into how best to design pro-poor interventions.
7. Data from villagers reflect a 'functional' village poverty pyramid, ranking the population by productivity of each livelihood group: the most productive at the top and the least productive at the bottom. The poverty-pyramid method of describing and monitoring trends in the structure of poverty at village level is described in Remenyi (1991, 1994, 2004).
8. Algebraically, if the bottom of the range is x_1 and the top of the range is x_2, then any number in between x_1 and x_2, defined as x_i, is given a number, y_i, within the limits of 1 and 5, using the following formula: $y_i = 5[x_2(x_2-x_i)/x_1]$, where $_i$ is a marker for any number in the poverty range for each indicator.

References

China State Council (2001) 'The Development-oriented Poverty Reduction Program for Rural China', White Paper, Beijing: Information Office China State Council.

Fan, S. (2002) 'Public Investment, Growth and Poverty Reduction in Rural China and India, Figure 1: Rural Poverty Incidence', Washington, DC: World Bank, available at http://poverty.worldbank.org/files/12402_SFan-Presentation.pdf (retrieved September 2007).

Henry, Carla, Manohar Sharma, Cecile Lapenu, and Manfred Zeller (2003) 'Microfinance Poverty Assessment Tool, International Food Policy Research Institute', *Technical Tools Series* No. 5, Washington, DC: CGAP.

Leading Group Office for Poverty (LGOP) (2000) 'Key Poverty Reduction Documents of the Chinese Government, PRC', paper prepared for the International Conference on China's Poverty Reduction Strategy in Early 21st Century, Beijing, 16 May.

Leading Group Office for Poverty (LGOP) (2002) 'Guidelines for Rural Poverty Alleviation and Development in China 2001–2010', Beijing: China State Council.

Leading Group Office for Poverty (LGOP), United Nations Development Programme (UNDP), and World Bank (2000) 'China, Overcoming Rural Poverty', Report No. 21105-CHA, Washington, DC: Rural Development and Natural Resources Unit, East Asia and Pacific Region, World Bank.

Li, Xiaoyun, Zhou Li, Yanli Liu, Yonggong Liu, Joe Remenyi, Sibin Wang, and Chungtai Zhang (2002) 'CPAP, A Methodology for Participatory Development Planning for Poverty Reduction in China', Beijing: Asian Development Bank, available at www.adb.org/prcm (retrieved September 2007).

Li, Xiaoyun, Wang Goliang, Joe Remenyi, and Pam Thomas (2003) *Training Manual for Poverty Analysis and Participatory Planning for Poverty Reduction* (in Chinese), Beijing: China International Books.

Ravallion, Martin and Shaohua Chen (2007) 'China's (uneven) progress against poverty', *Journal of Development Economics* 82: 1–42.

Remenyi, J. V. (1991) *Where Credit is Due: A Study of Credit Based Income Generation Programmes for the Poor in Developing Countries*, London: IT Publications.

Remenyi, Joe (1994) 'Poverty targeting' (chapter 8) in Bill Geddes, Jenny Hughes, and Joe Remenyi (eds.), *Anthropology and Third World Development*, Geelong: Deakin University Press.

Remenyi, Joe (2004) 'The poverty of development', in Damien Kingsbury, Joe Remenyi, John McKay, and Janet Hunt (eds.), *Issues in Development: Inclusiveness, Governance and Progress in Thinking on Development*, London: Palgrave Macmillan.

Remenyi, Joe and Max Kelly (2007) 'Own-poverty evaluation network – an *OPEN* approach to poverty assessment and monitoring for sustainable pro-poor development', in Clem Tisdell (ed.) *Poverty, Poverty Alleviation and Social Disadvantage: Analysis, Case Studies and Policies* (Volume 1, Chapter 7), New Delhi: Serials Publications.

Remenyi, Joe and Xiaoyun Li (2004) 'Towards sustainable village poverty reduction? The development of the CPAP approach', in John Taylor and Jannelle Plummer (eds.) *Capacity Building in China*, London: DFID.

United Nations (UN) (2006) 'Millennium Development Goals: China's Progress', Second MDG Report, Beijing: UN Resident Co-ordinator.

USAID (2004) 'Developing Poverty Assessment Tools Project, Poverty Assessment Tools', available at http://www.povertytools.org (retrieved September 2007).

World Bank (2001a) *Poverty Trends and Voices of the Poor*, Washington, DC: World Bank.

World Bank (2001b) *China: Overcoming Rural Poverty*, Washington, DC: World Bank.

Participatory risk assessment: a new approach for safer food in vulnerable African communities

Delia Grace, Tom Randolph, Janice Olawoye, Morenike Dipelou, and Erastus Kang'ethe

Women play the major role in food supply in developing countries, but too often their ability to feed their families properly is compromised; the result is high levels of food-borne disease and consequent limited access to higher-value markets. We argue that risk-based approaches – current best practice for managing food safety in developed countries – require adaptation to the difficult context of informal markets. We suggest participatory research and gender analysis as boundary-spanning mechanisms, bringing communities and food-safety implementers together to analyse food-safety problems and develop workable solutions. Examples show how these methodologies can contribute to operationalising risk-based approaches in urban settings and to the development of a new approach to assessing and managing food safety in poor countries, which we call 'participatory risk analysis'.

Background

Millions of small-scale farmers in Africa, many of them women, efficiently supply the surging demand for livestock products. Most meat, milk, eggs, and fish is sold in informal markets where conventional regulation and inspection methods have failed and where private- or civil-sector alternatives have not emerged. Studies that look for disease in these markets inevitably find it; the corollary is an enormous burden of sickness borne by poor consumers, as well as blocked access for poor farmers to emerging higher-value outlets such as supermarkets.

In response to the problem of unsafe food in informal markets, the International Livestock Research Institute (ILRI), an Africa-based international research institute, and its partners started to conduct research on urban livestock value chains. Research projects in urban settings in Uganda, Kenya, and Nigeria (Mwiine 2004; Kang'ethe 2005; Olawoye 2006) had the common objective of better understanding the benefits and harms of livestock keeping and better managing the health risks associated with livestock keeping. These projects used an eclectic mix of participatory methods, gender analysis, household questionnaires, collection of epidemiological data, and laboratory analysis, which were applied to various degrees and with various levels of success to the problem of food safety.

In this report of work in progress, we give some background and context, followed by a flavour of results from the gender, participation, and human health interface; we then discuss lessons learned in the application of participatory methods and gender analysis to the problem of food safety; finally, we describe how participatory methodologies and gender analysis are being incorporated into a new approach for assessing and managing health risks associated with livestock, an approach that we are calling 'participatory risk analysis'.

Poverty and gender inequities impair women's ability to feed their families safely

Women are the caretakers of the food supply in developing countries, but too often their ability to feed their families properly is compromised by lack of physical and human capital. One of the consequences of poverty and ignorance is high levels of food-borne disease. Biological contamination causes 2 billion illness episodes annually, with as much as 70 per cent of diarrhoea episodes among the under-fives being linked to biologically contaminated food.

Food-borne disease is a problem in rich as well as poor countries, and there is widespread consensus that risk analysis is the best way of managing it. Risk analysis has three components: risk assessment, risk management, and risk communication (Figure 1). The first step, and subject of this report, is risk assessment, which provides both an estimate of harm and the probability of harm occurring. It offers a science-based, structured, transparent method for answering the questions that matter to policy makers and public alike: is this food safe? Is the risk big and important? What efforts are appropriate to reduce the risk? To be useful, risk assessment must be followed by action to mitigate those risks that are unacceptable to stakeholders. Risk management uses pathway approaches (from stable to table) and probabilistic modelling to identify critical control points and apply strategies to remove or minimise risk. The third component, integral to risk analysis, is risk communication: the iterative process of communicating risk to those affected by it and incorporating their feedback into risk assessment and management.

Why is risk assessment not used in developing-country domestic markets?

Although risk assessment is current best practice and the keystone of international trade, its use in developing countries has been limited. In particular, it has not been applied to the domestic markets where most poor people sell and buy food, yet where levels of hygiene and safety are lowest, and vulnerability to food-borne disease highest. This is understandable, given the origin of risk assessment in the very different context of food production in developed countries. These systems tend to be large-scale, high-volume, mechanised, standardised,

Figure 1:   Risk analysis comprises risk assessment, management, and communication

and documented, while developing countries have diverse, non-linear, shifting, and data-scarce systems in which formal and informal (or traditional) food-supply systems co-exist and overlap; views of various stakeholders on food-safety objectives diverge; there is low consumer willingness or ability to pay for improved food quality, and low enforcement capacity. It is hardly surprising that radical adaptation is needed for risk-based approaches to work in such environments.

Recent analysis of research that successfully transforms knowledge into action suggests the importance of 'boundary-spanning actions' which create new spaces where different stake-holders can work together on joint activities or products. Participatory research and gender analysis can be one such bridge. Since their introduction in the 1970s, participatory methods and techniques have become central to community development. They are promoted on the basis that they are (compared with non-participatory methods) more effective, more sustainable, less costly, and more ethical in their inclusion of the poor in the planning and decisions that affect them (Karl 2000). In the field of animal health, participatory methods have been success-fully applied to community-based animal-health programmes, participatory epidemiology, and participatory disease surveillance and response (Jost *et al.* 2007). The gender-differentiated nature of food preparation, production, and obtaining food warrants the application of a gender perspective to participatory analysis of food safety.

Experience in applying risk-based approaches

Since the mid-1990s, a number of food-safety projects have attempted to incorporate participatory methodologies and gender analysis. A cross-cutting conclusion from our studies was that participatory methods were easier and less costly than conventional surveys, yet they generated comparable results and much higher levels of participation and research satisfaction. Another conclusion was that strong differences in access, power, and exposure to risk in the area of live-stock keeping necessitate the collection of gender-disaggregated data and separate analyses with women's and men's groups. Table 1 gives examples of the insights obtained by participatory and gender analysis in Nairobi, Kampala, and Ibadan, and the unique contribution that they can make to food-safety research.

Of course, the socio-cultural context has a major influence. For example, we found that while in all countries the majority of urban cattle farmers were women, men and women in East Africa kept similar numbers of cattle, and in Nigeria men owned ten times more cattle than women. In Kampala, the wealthier a woman was, the more cattle she kept; the wealthier a man was, the fewer cattle he kept. In Nigeria, women from Muslim communities had a minimal role in live-stock trade, but in Christian communities participation was extensive. Gender-disaggregated data also revealed a differential perception of involvement in livestock keeping. In Nairobi, for instance, women gave men lower scores for involvement in all dairy activities, but when the men did the scoring for the same activities, they gave themselves higher scores.

Perhaps the most important application of these methodologies is in understanding risk exposure. Participatory methods are excellent for mapping and describing dynamic processes, while the marked gender difference in roles associated with livestock and food leads to an equally marked difference in risks (see Figure 2 for an example).

Despite these encouraging findings of the usefulness of participatory and gender analysis, two challenges emerged.

- Almost no syncretisation of quantitative and epidemiological analysis with participatory and gender analysis took place. Participatory and formal data collection remained separate and sequential, with little integration or cross-fertilisation.

Table 1: Contributions of participatory methodologies and gender analysis to food-safety research

Participatory methodologies provide rapid, cheap, adequate information on livestock keeping and processing	Gender analysis reveals women's different (often higher) exposure to food-borne disease (FBD)
• *Frequency and type of livestock kept.* Not only poultry are kept in these cities, but also large animals such as dairy cows (even camels) and niche species such as deer, rabbits, snails, and grass-cutters.†	• *Women's higher exposure to high-risk activities.* Women have greater involvement in feeding, milking, and cleaning of livestock than men. In Nairobi, for example, women cleaned the cow-shed 30 per cent more often than men and did milking 40 per cent more often.
• *Importance of livestock to livelihoods.* Livestock keeping is a major income-generating activity. In Kampala, for example, households keeping chickens derived more than one-third of their income from this.	• *Women and men exposed to different diseases.* Differential livestock ownership is common, with women having relatively greater involvement in poultry and small-ruminant keeping.
• *Motivation for keeping livestock.* In most areas, both men and women ranked income derived from livestock as more important than household consumption.	• *Women much more exposed to FBD.* Women predominate in food and by-product processing, food preparation, and selling of ready-to-eat food.
	• *Women have less exposure to some risks.* In Ibadan, women kill and process pigs and poultry, but not larger animals. Some men said they can never eat the meat of a cow or goat slaughtered by a woman.
Participatory methodologies reveal unusual practices and give insights which might be missed in conventional surveys	**Gender analysis reveals how different gender roles result in different incentives and capacity for risk management**
• *Risks unique to certain contexts.* In Ibadan, sale of dog-meat and gin was an important livelihood strategy for women. Risks from carnivore meat differ from those from omnivore meat and require different management strategies.	• *Different channels for reaching women.* Livestock keeping is typically a secondary activity. But, while for men it is more likely to be combined with crop growing or salaried work, for women it is more likely to be combined with petty trade.
• *Cultural beliefs can be a disincentive for improving animal health.* In Ibadan, one man reported that he had a pregnant goat that died. He was advised to process and eat the goat, but he refused, believing that if he did not accept the loss, a pregnant woman in his household might die.	• *High access but low control is a recipe for unmanaged risk.* Women do not always have the power to make decisions over selling animals or their products (even in the case of poultry). In Nairobi, women had control over sale of milk, but not sale of cattle.
• *A wide range of indigenous practices exists.* In Kampala, an average of 17 strategies were used that effectively improve food quality and mitigate the risk of FBD.	• *Prioritisation reflects division of labour.* In Nigeria, where women are generally responsible for feeding and watering the animals, but men are responsible for providing the feeds, men's groups ranked lack of feed higher than women's groups did.

† Grass-cutter (also known as Savannah Cane-rat, *Thryonomys swinderianus*) is a rodent native to Africa.

	Women	Men
4:00 to 4:30	Wake up	
	Make a fire	
	Boil water for washing udders	
4:30 to 5:30	Milk cow* or supervise hired workers	
	Sieve*, weigh*, and pack* milk for sale	
	Deliver direct to customers or traders	
5:30 to 7:00	Prepare breakfast	Wake up
	Wake children and help young ones prepare for school	Wash
		Eat breakfast
	Clean milk cans* and breakfast utensils	
7:00 to 9:30	Clean cattle shed*	Treat sick cattle*
	Feed cattle*	Fetch or supervise fetching of
	Fetch fodder and water	fodder
	Attend to other livestock*	
	Wash clothes	
9:30 to 11:30	Participate in outside activities – church, women's groups, etc.	Work
11:30 to 1:30	Prepare lunch	If working close by, return for lunch
	Serve lunch	
	Wash up	
1:30 to 2:30	Give fodder and water to cattle*	Rest
2:30 to 4:00	Work in kitchen garden*	Socialise
4:00 to 5:30	Milk cow* or supervise hired workers	Supervise milking
	Sieve,* weigh,* and pack* milk for sale	
	Deliver milk direct to customers or traders	
5:30 to 11:00	Clean and wash house	
11:00	Retire to bed	Retire to bed

* High exposure to risk of zoonotic disease.

Source: Kang'ethe *et al.* (2005)

Figure 2: Daily activities for women and men keeping dairy cattle in Dagoretti, Nairobi

- Although risk-based approaches were used, and risk assessments planned, none of the projects succeeded in generating the type of data that could be used for a formal risk assessment, which is the basis of risk management. In the absence of a formal risk assessment, the studies were not able to introduce evidence-based strategies for risk mitigation.

In order to overcome these challenges, ILRI and its partners are developing a framework and tools to facilitate the integration of participatory methods and gender analysis into the assessment and management of risks associated with food.

Incorporating participatory tools and gender analysis

Our experiences suggest that operationalisation of risk-based approaches to food-safety problems in informal markets requires structured support which provides examples, road-maps, and tools drawn from participatory research and gender analysis. We propose a conceptual structure based on the risk-analysis framework of the *Codex Alimentarius* (CA), a body established by the WHO and FAO as a global reference for food-safety issues. The CA framework divides risk assessment into four linked steps (FAO and WHO 2003):

1. Hazard identification – *What is present? Is it harmful?*
2. Hazard characterisation – *What harms are caused? How does harm depend on dose?*
3. Exposure assessment – *How does the problem get from source to victim? What happens to it along the way?*
4. Risk characterisation – *What is the magnitude and likelihood of harm?*

For each of these steps, we suggest a series of participatory and gender-analysis tools which can complement or even substitute for conventional analysis. Examples of appropriate tools are shown in Figure 3. Details of the livestock tools can be found in Kirsopp-Reed (1994).

When there are no pre-existing data, or where cost is an over-riding constraint, these methods can be used in conjunction with literature reviews to construct a first-pass risk assessment. Where resources allow and there is a need for quantitative estimates of risk, these methods can be combined with epidemiological surveys and probabilistic modelling, allowing a rapid, low-cost risk assessment which gives information that is 'good enough' for decision support and risk management. This model is being field-tested in a project working with women's

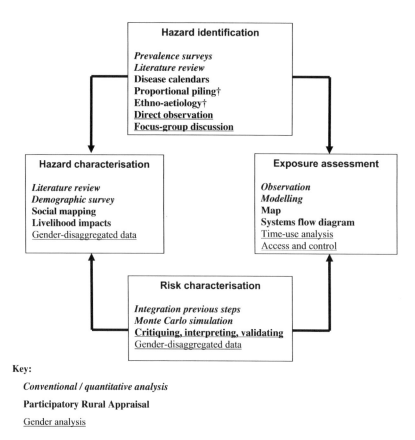

Figure 3: Integration of participatory research and gender analysis into risk assessment

groups involved in the informal markets of Ibadan; we expect that it will be further refined as a result of this testing.

Conclusion

Safer animal-source food can generate both health and wealth for the poor, but attaining safe food and safe food production in developing countries requires a radical change in food-safety system management, scope, focus, and instruments. Risk-based approaches that look at the probability and extent of harm caused to consumers, rather than the presence of pathogens, are current best practice, but have not been shown to work in the heterogeneous, complex, data-poor informal markets where most of the poor buy and sell. Integrating risk assessment with participatory methodologies and gender analysis is a promising solution. Generating credible evidence using such an approach will be critical to better understanding and better managing food safety in developing countries.

References

Food and Agriculture Organization of the United Nations (FAO) and World Health Organization (WHO) (2003) *Assuring Food Safety and Quality: Guidelines for Strengthening National Food Control Systems*, Rome: FAO, and Geneva: WHO.

Jost, C. C., J. C. Mariner, P. L. Roeder, E. Sawitri, and G. J. Macgregor-Skinner (2007) 'Participatory epidemiology in disease surveillance and research', *Scientific and Technical Review* 26 (3): 537–47.

Kang'ethe, E. (2005) 'Characterization of Benefits and Health Risks Associated with Urban Smallholder Dairy Production in Dagoretti Division, Nairobi, Kenya', unpublished report, University of Nairobi.

Karl, M. (2000) *Monitoring and Evaluating Stakeholder Participation in Agriculture and Rural Development Projects: A Literature Review*, Rome: Food and Agricultural Organization, available at http://www.fao.org/sd/PPdirect/PPre0074.htm (retrieved 28 January 2008).

Kirsopp-Reed, K. (1994) 'A review of PRA methods for livestock research and development', *PRA Notes* 20 Special Issue on Livestock, London: IIED.

Mariner, J. (2000) *Manual on Participatory Epidemiology – Method for the Collection of Action-Oriented Epidemiological Intelligence*, Rome: Food and Agriculture Organization.

Mwiine, N.F. (2004) 'Benefits and Health Risks Associated with Milk and Cattle Raised in Urban and Peri-urban Areas of Kampala City', MSc thesis, Uganda: Makerere University.

Olawoye, J. (2006) 'Improving Benefits of Urban and Peri-urban Livestock Production Through Management of Associated Human and Environmental Health Risks in Nigeria', Nairobi: ILRI.

Pro-poor values in agricultural research management: PETRRA experiences in practice

Ahmad Salahuddin, Paul Van Mele, and Noel P. Magor

PETRRA was an agricultural research-management project which used a values-based approach in project design, planning, and implementation. Through an experiential learning process, agricultural research and development (R&D) institutes, NGOs, private agencies, and community-based organisations rediscovered and improved the understanding of their strengths in meeting development commitments. The project successfully showed how values-based research can meaningfully be implemented and a sustainable pro-poor impact achieved.

Introduction

Poverty Elimination through Rice Research Assistance (PETRRA) was a research project implemented in Bangladesh from April 1999 to August 2004. It operated with a budget of £9.5 million, funded by the UK Department for International Development (DFID), and managed by the International Rice Research Institute (IRRI) in close partnership with the Bangladesh Rice Research Institute (BRRI). The project aimed to enhance the livelihood security of poor farmers by increasing the production and productivity of rice-based farming systems through poverty-focused research. Rice was the entry point, and research was to support a strategy for poverty elimination. The project avoided targeting production; instead it targeted poor farmers and not the large producers. PETRRA started with people – resource-poor farm households – not technology (Orr and Magor 2007).

The project objective and logical framework statements were adjusted over time to reflect the praxis of being poverty-focused. PETRRA managed 45 sub-projects across more than 50 institutions and organisations which represented traditional international and national research institutions, government rural-development institutions, NGOs at national and local levels, the private sector, and community organisations. Many of the sub-projects comprised partnerships. The sub-projects covered technology development, uptake-methods research, and policy research and dialogue, with more than 700 research and development (R&D) people engaged. The project established partnerships of multiple actors with roles of either research or extension provision. For PETRRA, the challenge was to encourage a poverty focus and the processes related to such a focus. The concept of 'values-based research' grew out of the action–reflection practice that was used throughout the project (Quin *et al.* 2003). Research prioritisation began with a stakeholder analysis that was inclusive of resource-poor farmers, men and women, local and regional government and extension staff, and civil-society actors. This was a first step towards engaging farmers in identifying research issues and it opened up a more active role for the end-users in the entire development process. Overall, this illustrates a shift towards a more participatory R&D philosophy and towards a model based on multiple sources of innovation.

In the fourth year, communication was added to the logical framework as a key project output, for two reasons: (1) downward accountability – communicating innovation to the primary stakeholders, the farmers; and (2) upward accountability – communicating PETRRA management practices and project innovations to agricultural R&D organisations and policy makers. Communication emerged as a major factor in successful consolidation of the project outcomes and in ensuring greater impact.

The project management unit (PMU), donors, and reviewers were never completely happy with the formulations of the statements of the objectives (*super goal, goal, purpose*, and *outputs*) and their vertical and horizontal links and logics. This situation reflected the complexity and dynamic nature of the project and was seen as positive. Project stakeholders, including the donor, began to understand that they were far better placed than external consultants to make regular adjustments to the statements, since they were continuously learning. PETRRA project documents mentioned certain values as cross-cutting issues, but only after several iterations of practices and reviews did their links to the project objectives in the logframe structure become clear.

PETRRA's emerging values-based management approach

PETRRA identified, developed, and defined various cross-cutting issues, including poverty focus, demand-led research, participation, partnership, gender, linkage and network, and competition in research management. These formed the value base of the project and played a crucial role in conceptualising and developing PETRRA's agenda. Values were defined as central beliefs and purposes of the society – in this case the organisation or the project (Jary and Jary 1991). PETRRA strived for best practice in the following respects.

- Working with resource-poor farmers to address *poverty*.
- Conducting research as per *demand* and priority of the resource-poor farmers.
- Conducting, sharing, and evaluating research with both *men and women* members of resource-poor households.
- Conducting research that ensured *participation* of resource-poor men and women in all stages of the project cycle: planning, designing, implementation, monitoring, and evaluation.
- Conducting research by establishing appropriate and effective *partnership* of agencies to ensure the use of pro-poor technology, dissemination methods, and policy.
- Ensuring that research outputs were sustained through *linkage and network* development, with appropriate agencies, to ensure that the interests of the poor were represented.
- *Communicating* effectively with farmers and policy makers to disseminate, up-scale, and consolidate learning.
- Using a *competitive* process as a way of identifying competent suppliers of agricultural R&D to facilitate the achievement of pro-poor outcomes.

These practices evolved while working with poor households. At the same time, PETRRA established the definition, scope, concept, and practical means to translate the values into actions. All these elements together formed PETRRA's *values-based research approach*. Table 1 captures its systematic unfolding.

Tables 2(a) and 2(b) explore the link between the values and PETRRA outputs. In Table 1, 'communication' is mentioned as a value. It evolved during the project, finally being included as an output. In Table 2(b), 'competitive' is mentioned as a value that aimed to achieve the desired outcomes.

The values adopted by PETRRA are not new, and much literature is devoted to their useful-ness for developing a pro-poor agricultural research system. The significance of the PETRRA project was to identify the important values for a poverty focus through action and reflection with partners and then incorporate these into a management system that was coupled with capacity building to facilitate the process. The actors in each sub-project incorporated the values through action and then, as a collective of sub-projects, shared that experience.

Institutionalising values-based research

The scope of the initial proposal was rather limited, but that did not prevent the project from becoming innovative. In a way it allowed the project to blossom naturally. The donor and host agencies also allowed the project to evolve: they were not rigid, but rather appreciative of new ideas and innovations.

The PMU was open-minded and strove to be responsive to the needs of resource-poor farmers. It initiated ideas, included new outputs, adjusted project purpose, invited and enter-tained new ideas from project stakeholders and outsiders, reviewed suggestions, and reacted according to the situation. It also exercised the freedom to be neutral, even towards its own organisation, IRRI.

Although the project recognised and brought into practice various values, a lot could still be done to establish these values within agricultural research institutes like IRRI and BRRI, and to identify appropriate ways to institutionalise them in the overall R&D system. Some researchers emerged as 'champions' promoting values, but the extent to which this learning carried over into day-to-day work beyond the specific sub-project varied. To establish a culture that embraces values within a project is not enough: those values need to be embedded in the agencies, so that the praxis continues. International organisations like IRRI have the scope to influence national agri-cultural research systems (NARS) in rice-producing developing countries. Through PETRRA, IRRI showed that it can facilitate and establish an effective values-based research culture.

Various examples show PETRRA's continued impact. The World Bank and the International Fund for Agricultural Development (IFAD) decided to jointly support the National Agricultural Technology Project, with an estimated grant of US$ 84.5 million (BARC 2007). PETRRA values such as a competitive grant system, demand-led research and extension, poverty focus, and partnership are included. Immediately after the project ended, DFID agreed to estab-lish a *Projukti* (technology) Foundation in Bangladesh that would reflect PETRRA values – but not much progress was made, as the government did not agree to make it a body independent of government control.

Some project partners acquired grants from the CGIAR Challenge Program for Water and Food to follow up on their successful PETRRA research. The two projects provide IRRI with further experience to consolidate and internalise values-based research.

PETRRA's experiences with extension methods research were documented in the book *Inno-vations in Rural Extension – Case Studies from Bangladesh* (Van Mele *et al.* 2005). In his review of the book, Robert Chambers commented: 'if any donor agency is looking for a

Table 1: Changes in partners' perceptions about values within PETRRA

Value	Year 1	Year 2	Year 3	Year 4	Year 5
Poverty focus	Partners were not aware of it	Agreed but most were not aware of the rationale and approach	Started practising well-being analysis	Revised the portfolio of clients	Most clients were resource-poor
Demand-led	Sub-projects defined demands	Sub-projects referred to demands expressed from the stakeholder analysis done by PMU	Sub-projects conducted extended analysis to sharpen the demands of the resource-poor farmers	Integrated resource-poor farmers' demands into the project management cycle	Partners recognised demand as the basis for responsive research
Gender	Partners were confused about the importance	Agreed to be more inclusive of women; gave some training on post-harvest issues	Agreed to train women in, and discuss with them, all aspects of farming (not just post-harvest)	Appreciated the importance of women in all aspects of farming	Appreciated the concept of family approach and women accessing all aspects of knowledge
Participation	Partners were aware, but did not practise and often resisted	Agreed to take training	Started using the approach	Started to appreciate its importance	Agreed participatory approach as a guiding principle
Partnership	Partners uncommon	Reluctantly accepted the idea of forming partnership	Started to realise the advantage	Started to appreciate the importance	Agreed to sustain the relationship for future collaboration
Linkage and network	Partners hardly had any linkage among GO–NGO, or GO–PS, or NGO–R&D institute	Government policy and project facilitated the relationship	Started appreciating the advantages	Appreciated the importance for sustainability of the innovations and impact	Most have recognised the advantage and a few have institutionalised the relationship and signed MoU between organisations
Communica-tion	Partners never interacted with farmers; scientific papers were the only targets	Farmers asked for materials	Partners participated in communication fair and contributed to newsletters	Started to appreciate materials for farmers and secondary stakeholders	All sub-projects produced a set of materials for farmers and shared the pride

GO – government organisation; PS – private sector; MoU – memorandum of understanding.

Table 2(a): PETRRA outputs and their linkages with the values

Value	Technology	Communication	Uptake
Output statement	*Improved rice production technologies appropriate to poor farmers identified or developed and tested in collaboration with them, PETRRA sub-project partners, and PMU*	*PETRRA management practices and research findings effectively communicated to relevant organisations and individuals involved in agricultural research and extension, and to policy makers*	*Improved methods for effective uptake of technologies identified, pilot-tested, and recommendations for improved uptake pathways made by PETRRA's sub-project partners and PMU*
Poverty focus	Select appropriate clients to work with, i.e. poor farming households	Ensure that poverty remains as the main focus of all communication activities, no matter whether the materials are targeted at farmers, extension workers, scientists, or policy makers	Develop innovative and appropriate pro-poor uptake methods that may or may not be different from those of non-poor
Demand-led	Research priorities based on needs of the clients, not decided unilaterally by the researchers	Guarantee and monitor demand for materials from all levels; sometimes may need to create demand for tested materials	Identify gaps in the system and identify appropriate uptake methods; farmers' demand should be at the centre of the analysis, which needs to be compatible with the interests of the partners concerned
Gender	Work with both male and female members of the households	While developing, testing, and disseminating communication materials (e.g. leaflet, poster, video, fact sheet), engagement with both men and women is considered; sometimes specific attention and tools are needed for women	Ensure that uptake methods for technology dissemination take into account the interests of both men and women. Where the target is the household, both men and women should be involved; each can be involved separately, if that appears more appropriate
Participation	Ensure participation of poor men and women farmers in all stages of the project cycle	While developing, testing, and disseminating communication materials, participation of both men and women is ensured in all stages	Involve farmers (men and women) and stakeholders of all levels in the research–development process
Partnership	Ensure proper partnership that can effectively help to develop, disseminate, and sustain the technology	Ensure appropriate partnership is formed at all levels for developing, testing, and disseminating communication materials; ensure that resource-poor farmers and material developers become partners	Establish strategic partnerships based on comparative advantage to ensure development of, research on, and sustainability of uptake methods

(continued)

Table 2(a): Continued

Value	Technology	Communication	Uptake
Linkage and network	Establish linkage and network during the project and thereafter, to help eliminate structural and institutional barriers to technology adoption	Ensure that partnerships are not lost once project ends, with ability to expand the social capital for potential future investments	Design a sustainable linkage and network as part of the research on uptake methods; this should not be threatened to be discontinued immediately after the project ends
Competitive	Commission most research on a competitive basis to identify competent suppliers; create level playing field through open bidding	Select partners for development and dissemination on competitive basis	Gather ideas from different suppliers of research through competition; uptake-methods research requires a series of facilitated discussions to develop and articulate research outlines

PMU – Project Management Unit.

Table 2(b): PETRRA outputs and their linkages with the values

Value	Capacity	Policy	Pro-poor model
Output statement	*Capacity of rice research system to undertake demand-led research sustainably enhanced*	*Key policy constraints to improved rice-dependent livelihoods identified and recommendations presented in key policy forums by PETRRA policy-research partners*	*A pilot model of an effective pro-poor competitive rice-research management scheme established and managed effectively by PMU*
Poverty focus	Train/orient researchers in different ways and means of poverty-focused research	Ensure that poverty issues are central to any policy research agenda	Ensure that poverty focus remains as the key value
Demand-led	Train/orient researchers in different approaches and techniques to identify poor (men and women) farmers' demands	Identify policy researchable issues with poor men and women farmers; avoid a top–down agenda	Monitor and adjust the model's relevance to ensure that it remains demand-led
Gender	Train/orient researchers in approaches and dimensions of gender-balanced research	Identify context of both men and women in policy research and formulate recommendations for both men and women	Ensure that gender-awareness is a strong component in the model

(*continued*)

Table 2(b): Continued

Value	Capacity	Policy	Pro-poor model
Participation	Train/orient researchers in approaches and techniques of participatory research in all stages of research	Involve in research the people who are affected by policy issues; concerned stakeholders should be involved during research and policy dialogues	Ensure that participation becomes the culture of the model
Partnership	Train/orient researchers in approaches and advantages of partnership for conducting demand-led participatory research	Involve all stakeholder levels (farmer, field worker, *upazila*/district, and national) in policy research, as national-level stakeholders are not able to represent all	Ensure that the model finds its strength in partnerships
Linkage and network	Train/orient researchers in approaches and inform them about the advantages of linkage and network to sustain the technology among its users	Establish linkage and network for continued follow-up and policy dialogue for sustainability	Ensure that the model always advocates linkages and networks to strengthen and sustain the model itself
Competitive	Train/orient researchers to equip them to participate in the competitive bidding system and be successful	Conduct policy research on a competitive basis by NGOs, community-based organisations, local government, private sector, media, and whoever is working with or for resource-poor farmers	Ensure that the model is developed, tested, and sustained through a competitive process and is exposed to competition

PMU – Project Management Unit.

cost-effective investment, it would be hard to do better than to provide the means to make this book cheap and accessible, and to send a great many copies with a covering letter to those concerned with agricultural research and extension policy and practice around the world' (Chambers 2007: 36). The 'Focal Area' concept developed during PETRRA is now commonly used by government and non-government organisations to jointly address poverty in northern Bangladesh. Partners formerly involved in extension-methods research continue to expand their activities in-country (for example, farmer-oriented seed models) and across South Asia (for example, women-oriented video production). BRRI continues to develop the Bangladeshi version of the rice knowledge bank (www.knowledgebank-brri.org), which is focused on semi-literate farmers and extension workers. These are but a few examples.

Conclusion

PETRRA offered an opportunity to experiment with socio-technical and institutional innovations for the development of a pro-poor agricultural research system. It created a lot of enthusiasm among its partners and wider stakeholders. Although its implementation efficiency was criticised, this was the price paid for operating in a mode of experiential learning. PETRRA successfully demonstrated the means by which such an endeavour can be shaped.

References

BARC (2007) 'Development Project Proposal (DPP) for National Agricultural Technology Project (NATP)', Dhaka: BARC, Ministry of Agriculture, Government of the People's Republic of Bangladesh.

Chambers, R. (2007) 'Book review: Innovations in rural extension', in A. Barclay (ed.) *Rice Today*, Los Baños, The Philippines: International Rice Research Institute.

Jary, D. and J. Jary (1991) *Collins Dictionary of Sociology*, Glasgow: HarperCollins.

Orr, A. and N. P. Magor (2007) 'PETRRA Project Strategy', in N. P. Magor, A. Salahuddin, T. K.Biswas, M. Haque, and M. Bannerman (eds.) 'PETRRA – An Experiment in Pro-poor Agricultural Research', Strategy Brief No. 2.1, Dhaka (Bangladesh): Poverty Elimination through Rice Research Assistance Project, International Rice Research Institute.

Quin, F., B. Musillo, K. Kar, and Z. Alam (2003) 'Fourth Output to Purpose Review Report, Poverty Elimination Through Rice Research Assistance (PETRRA) Project', Dhaka: Rural Livelihoods Evaluation Partnership.

Van Mele, P., A. Salahuddin, and N. P. Magor (eds.) (2005) *Innovations in Rural Extension: Case Studies from Bangladesh*, Wallingford: CABI Publishing.

Operationalising participatory research and farmer-to-farmer extension: the *Kamayoq* in Peru

Jon Hellin and John Dixon

While rural poverty is endemic in the Andean region, structural adjustment programmes have led to a dismemberment of agricultural research and extension services so that they are unable to serve the needs of smallholder farmers. The NGO Practical Action has been working in the Andes to address farmers' veterinary and agriculture needs. The work has included the training of farmer-to-farmer extension agents, known locally as Kamayoq. *The* Kamayoq *have encouraged farmer participatory research, and local farmers pay them for their veterinary and crop advisory services in cash or in kind. The* Kamayoq *model is largely an unsubsidised approach to the provision of appropriate technical services and encouragement of farmer participation. The model also illustrates that, in the context of encouraging farmer participation and innovation, NGOs have advantages over research organisations because of their long-term presence, ability to establish trust with local farmers, and their emphasis on social and community processes.*

Introduction

The use of participatory approaches is one way of enhancing rural innovation capacity, whether through increasing accessibility of externally developed technology, or jointly developing relevant and appropriate technology by farmers and scientists, or enhancing local capacity to address problems and devise solutions for them.[1] The importance of farmer participation in formal agricultural research became widely accepted in the 1980s and 1990s (for example, Bunch 1982; Scoones and Thompson 1994).

Participatory research can be divided into 'functional' and 'empowerment' purposes (Lilja and Bellon 2006). Functional purposes involve increasing the validity, accuracy, and particularly the efficiency of the research process and its outputs. They largely address the technologies that scientists should develop – for example, assisting plant breeders to identify the farmer-valued traits that breeding programmes should focus on. Meanwhile, empowerment purposes include strengthening farmers' capacity to analyse opportunities and set priorities for change and innovation.

Both the functional and empowerment aspects of participatory research are important. The former tend to be the domain of a wide variety of organisations, including research and development organisations, even though knowledge, information, and technology may be generated from a central source (such as a publicly funded research entity) and the information flow is linear (from researchers to farmers via extension agents) (Biggs 1990), but more often from multiple-source innovation networks. The danger is that mono-disciplinary theoretical recommendations may be made for what are, in fact, multi-faceted problems embedded in complex local agro-ecosystems and socio-cultural systems.

Furthermore, individuals participate in social change not as passive subjects, but rather as social actors whose strategies and interactions shape the outcome of development within the limits of the information and resources available. Active farmer participation is widely recognised as one of the key components of rural development (Chambers 1997). The confidence that comes from participation increases farmers' ability to learn and experiment. It is, therefore, critical that farmers are fully involved in the research process in order to ensure (as much as possible) that the technology complements their particular situation and that farmers are suitably empowered (Hellin *et al.* 2008).

There are, however, several challenges when it comes to operationalising participatory research, not least agreeing on the most cost-effective way forward and the most appropriate roles for different organisations such as research institutions, NGOs, and the private sector. These issues are highly pertinent in an age of structural adjustments that have led to a breakdown of classical publicly funded agricultural research and extension services, to the extent that these services are now unable to address the needs of farmers living in marginal environments. Private research and extension provision was expected to replace that previously provided by government. Few resource-poor farmers, however, are able to pay for this service and, as a result, it has generally been directed at larger commercial farmers (Chapman and Tripp 2003). There are, however, a number of initiatives that have encouraged farmer innovation and experimentation. One example is the *Kamayoq* in the highlands of Peru.

The *Kamayoq*

The provision of publicly funded research and extension in Peru has been much reduced since the mid-1980s. For example, the government agricultural extension programme run by what is now the *Instituto Nacional de Investigación Agraria* (INIA) employed 1400 extension officers in 1986, but fewer than 100 officers in 1992. Since the 1990s, Practical Action (formerly known as ITDG), a development NGO, has been working in Quechua-speaking farming communities in the Peruvian Andes to try to fill this void. Initially, the focus was on communities living in the valleys above 3500 metres, where common crops are maize, potatoes, and beans, and where families also have one or two head of cattle each, some sheep, and a number of guinea-pigs (a food staple in the Andes). Since 2003, the focus has broadened to include communities living above 4000 metres, where livelihoods depend on a combination of alpaca raising and potato cultivation.

For over 500 years, the Quechua, like most Latin American indigenous peoples, have been undervalued and marginalised. Influenced by the pedagogic approach of the Brazilian educator Paulo Freire, Practical Action (PA) recognised that one of the most effective ways to address farmers' needs was through a farmer-to-farmer extension approach that also encouraged farmer experimentation and learning by doing. In the early 1990s, PA began to train a number of farmer extension agents, known locally as *Kamayoq*, focusing initially on irrigation techniques.

Practical Action soon recognised that farmers' needs could best be met by broadening the focus beyond irrigation. In 1996, PA established a *Kamayoq* school in Sicuani, 140 km north of the city of Cusco. The objective was to train local farmers who would then be responsible for training other villagers and encouraging farmer experimentation. Up to the end of 2007, approximately 200 *Kamayoq* had been trained, of whom 20 per cent were women. Training courses at the school take place over an eight-month period both at the school and in different field locations, so that the *Kamayoq* can 'learn by doing' (de la Torre 2004).

Instructors at the school include staff from Practical Action, long-serving *Kamayoq*, and experts from regional universities in the cities of Puno and Cusco. During the training, the *Kamayoq* also visit INIA's experimental stations, other NGOs working in the region, and large-scale farmers. Throughout their training, the *Kamayoq* establish contact with technical experts from the private and public sectors, and with other farmers – a useful network which they can tap into when they need information and technical advice once they finish their training. This 'social capital' is recognised by many as one of the greatest benefits of the whole course.

Themes covered in the training meet local farmers' expressed needs and include irrigation, Andean crops, horticulture, livestock, forestry, agro-industry, and marketing. The *Kamayoq* do not become promoters of off-the-shelf technologies. On the contrary, the objective is to encourage the *Kamayoq* to work with farmers to generate local innovations in response to local agricultural and veterinary problems. This is important not just for empowering farmers, but also because farming conditions in the Andes are so complex and diverse that it is difficult to find a ready-to-use technology that can be adopted by a large number of farmers without adaptation. The *Kamayoq* are encouraged to see themselves as key players in a two-way flow of information: from the individuals and institutions promoting development, and from the local farmers to these same individuals and organisations. Indeed, in another context they would be considered as facilitators of the local innovations systems.

Combining participatory research and farmer-to-farmer extension

The *Kamayoq* have encouraged farmer participatory research. Successful initiatives have included the treatment of a maize fungus disease; the control of mildew on onions; and treatment of animal diseases. The most sought-after service is the last of these: the diagnosis and treatment of various animal diseases. One of the biggest problems in sheep and cattle (and increasingly alpaca) in the Andes is the parasitic disease *Fasciola hepatica*, commonly known as 'sheep liver fluke'. Although *F. hepatica* rarely kills animals, it does incapacitate them. Additionally, sick animals often weigh one third less than healthy ones, and infected bulls sell for around US$ 70 – some 65 per cent of the price of healthy bulls. In the case of cows, infection causes a reduction of more than 50 per cent in milk production. Weakened animals are also susceptible to a number of secondary diseases.

Few farm families could afford conventional medicines to control the disease. The discovery of a natural medicine to treat and control *F. hepatica* depended on a process of participatory research and development guided by the *Kamayoq*. A natural cure for *F. hepatica* in sheep had been discovered earlier by a local farmer who later became a *Kamayoq*. Between 1998 and 2000, the same farmer, along with Practical Action, national researchers, and local villagers, experimented with a cure for *F. hepatica* in cattle. Farmers played a direct and active role throughout. Farmers focused on a number of plants that were known to have medicinal properties. They tested medicines made from different combinations of these plants on their own infected animals. Experiments were designed to ensure that any treatment could subsequently be easily prepared and administered by the farmers themselves. The medicine, which contains garlic and artichoke, is administered to the animals in oral form. Farmers are now involved in experiments to find a cure for *F. hepatica* in alpacas.

The natural medicine is cheaper than conventional medicines: the cost of treating a sick animal with conventional medicine is approximately US$ 2.50 per animal; in the case of the natural medicine, it is US$ 0.60 per animal. More than 3000 families now use the natural medicine for controlling *F. hepatica* in the highland provinces near to Sicuani, and villagers have treated approximately 30,000 cattle and 7000 sheep (Hellin *et al.* 2005). The medicine's widespread use has led to fewer sick animals, higher milk yields, and diversification into a range of milk products including yoghurt and cheese, as well as the cultivation of 'new' crops such as onion and carrots.

Scaling up

The school is not expensive to run, and in some cases the *Kamayoq* are able to pay for part of their training. Still, it is unrealistic at this stage to expect them to cover more than a small percentage, so the continued success of this development initiative requires external funding; although it is not unrealistic to work towards a system whereby the *Kamyoq* will eventually be able to pay for their own training. The key to the success of the *Kamayoq* model is that local farmers highly value the assistance provided by their fellow *Kamayoq* and are willing and able to pay for this assistance. Farmers pay the *Kamayoq* for their services in cash, in kind, or in the promise of future help through an indigenous system known as '*ayni*' (Hellin *et al.* 2006). It is largely an unsubsidised farmer-to-farmer extension service. The *Kamayoq* do not receive any particular additional service such as preferential access to technology, seeds, or finance, although they do of course tend to have the knowledge and confidence to adopt technology on their own farms.

It is farmers' willingness to pay that makes the *Kamayoq* model so interesting. In general, where technical services are provided in response to demand, providers tend to serve only the better-off farmers and to ignore those living in marginal areas (Miehlbradt and McVay 2005). This is because private service providers tend to believe that only meagre profits (if any) can be made from working in these often remote areas. The *Kamayoq* model demonstrates that the payment-for-services approach is possible in more marginal areas.

The success of the *Kamayoq* model can be seen at different levels, including increased and/or sustained demand for the *Kamayoq's* technical advice; the demand from *Kamayoq* for refresher and new training; and interest on the part of other organisations in replicating the *Kamayoq* approach – the *Kamayoq* are increasingly being contracted by public and private organisations to extend the model beyond the communities and region where they have operated to date. Meanwhile, Practical Action has extended the *Kamyoq's* activities to higher-altitude areas, where farmers' livelihoods depend on alpaca and native potato varieties.

Like most conventional agricultural extension provision, the *Kamayoq* have worked predominantly on improving and increasing production at the farm level. The next step is, therefore, to determine how the *Kamayoq* model could be developed to provide farmers with the business services that they need in order to benefit from emerging market opportunities (for example, market linkages and processing skills), and to encourage farmer innovation in order to enhance this market access. The second author of this article was involved in just such an initiative in a remote and poor mountain community in Nepal. Here, farmers were empowered through, *inter alia*, training in social entrepreneurship, micro-finance, and market analysis, through the Farmer-Centred Agricultural Resource Management (FARM) programme (Dixon 1996). It was noticeable that the technical training empowered the women farmers to the extent that they actively engaged in policy dialogue (a delegation visited the prime minister). Meanwhile the men searched for innovations: a farmer group funded one member to travel half way across Nepal to obtain better-tasting, higher-price ginger, which was brought back and cultivated successfully in the village, and subsequently sold in local markets.

Conclusions

There are challenges to operationalising farmer participatory research, especially in the context of debilitated publicly funded research and extension provision. While the functional aspects of participatory research are best carried out by research organisations, these same organisations' direct impact on their empowerment is often limited to relatively few farmers. This is because most participatory research initiatives carried out by research organisations do not have sufficient presence on the ground, and do not involve the required interaction with farmers, to generate and support direct empowerment of large numbers of farmers.

The achievement of this would necessitate a longer-term and more direct interaction with farmers than that usually associated with the *modus operandi* of a research organisation (that is, projects of three to five years). In addition, the impacts of most participatory research (carried out by research organisations) on farmers' innovation capacity and livelihoods are seldom sufficient, in themselves, to justify the expenditure of the research process.

The *Kamayoq* model in Peru suggests that farmer empowerment *per se* is best carried out by development organisations whose longer-term interaction with farmers is likely to ensure that greater numbers of farmers benefit. The success of the *Kamayoq* model has partly depended on the creation of a market for technical advisory services. This, in turn, has ensured a supply of competent advisers (the *Kamayoq*) and the stimulation of a demand for advisory services as farmers become aware of new opportunities and the incentives for their own experimentation and innovation.

Note

1. This article builds on material published in Hellin *et al.* (2005, 2006).

References

Biggs, S. (1990) 'A multiple source of innovation model of agricultural research and technology promotion', *World Development* 18 (11): 1481–99.

Bunch, R. (1982) *Two Ears of Corn: A Guide to People-Centered Agriculture*, Oklahoma, OK: World Neighbors.

Chambers, R. (1997) *Whose Reality Counts? Putting the First Last*, London: IT Publications.

Chapman, R. and R. Tripp (2003) 'Changing Incentives for Agricultural Extension: A Review of Privatized Extension in Practice', Agriculture Research & Extension Network Paper No. 132, London: Overseas Development Institute.

de la Torre, C. (2004) *Kamayoq: promotores campesinos de innovaciones tecnológicas*, Lima, Peru: ITDG Latin America.

Dixon, J. (1996) 'Institutionalizing farmer-centred sustainable agricultural resource management', *Journal of Farming Systems Research and Extension* 6 (1): 93–107.

Hellin, J., D. Rodriguez, J. Coello, W. Chañi, and A. Tayro (2005) 'The Kamayoq in Peru: combining farmer-to-farmer extension and farmer experimentation', in J. Gonsalves, T. Becker, A. Braun, D. Campilan, H. De Chavez, E. Fajber, M. Kapiriri, J. Rivaca-Caminade, and R. Vernooy (eds.) *Participatory Research and Development for Sustainable Agriculture and Natural Resource Management: A Sourcebook*, Laguna, The Philippines: International Potato Center – Users' Perspectives with Agricultural Research and Development, and Ottawa, Canada: IDRC.

Hellin, J., C. de la Torre, J. Coello, and D. Rodríguez (2006) 'The Kamayoq in Peru: farmer-to-farmer extension and experimentation', *LEISA magazine* 22 (3): 32–4.

Hellin, J., M. R. Bellon, L. Badstue, J. Dixon, and R. La Rovere (2008) 'Increasing the impacts of participatory research', *Experimental Agriculture* 44: 81–95.

Lilja, Nina and Mauricio Bellon (2006) *Analysis of Participatory Research Projects in the International Maize and Wheat Improvement Center*, Mexico, DF: CIMMYT.

Miehlbradt, A. and M. McVay (2005) 'From BDS to Making Markets Work for the Poor', Geneva: ILO.

Scoones, I. and J. Thompson (eds.) (1994) *Beyond Farmer First: Rural People's Knowledge, Agricultural Research and Extension Practice*, London: IT Publications.

Using community indicators for evaluating research and development programmes: experiences from Malawi

Jemimah Njuki, Mariam Mapila, Susan Kaaria, and Tennyson Magombo

Evaluations involving stakeholders include collaborative evaluation, participatory evaluation, development evaluation, and empowerment evaluation – distinguished by the degree and depth of involvement of local stakeholders or programme participants in the evaluation process. In community participatory monitoring and evaluation (PM&E), communities agree programme objectives and develop local indicators for tracking and evaluating change. PM&E is not without limitations, one being that community indicators are highly specific and localised, which limits wide application of common community indicators for evaluating programmes that span social and geographic space. We developed community indicators with six farming communities in Malawi to evaluate a community development project. To apply the indicators across the six communities, we aggregated them and used a Likert scale and scores to assess communities' perceptions of the extent to which the project had achieved its objectives. We analysed the data using a comparison of means to compare indicators across communities and by gender.

Introduction and background

Community-driven participatory monitoring and evaluation (PM&E) approaches build on the concepts and ideas developed by the Institute of Development Studies (IDS) at the University of Sussex (Guijt and Gaventa 1998). Community PM&E is an instrument to support systematic reflection (Germann and Gohl 1996), learning (Guijt and Gaventa 1998; McAllister 1999), the generation of knowledge for decision making (McAlister 1999), and process-oriented management at the community level (Probst 2002). This type of PM&E approach is unique because of the emphasis on developing a system that is managed and supported by local communities, for their own purposes. In community-driven PM&E, community members themselves identify their own objectives and initiate activities to achieve those objectives. They develop their indicators for measuring progress towards achievement of the objectives, are in charge of the data collection and analysis, and use the PM&E results to adjust their activities. The (local) indicators are based on the experiences, perceptions, and knowledge of the local people. The purpose of community-driven PM&E is to empower the local community to initiate control and take corrective action, and to empower them to improve their social well-being.

Approaches for developing monitoring and evaluation systems and monitoring indicators are wide and varied. They range from engaging expert evaluators to determine what indicators are appropriate and relevant, to employing participatory processes that involve communities and local stakeholders. Participatory processes themselves are varied, and the term has been used to describe very different levels of community participation and control over the process. Participation in evaluation spans a gradient from complete community-controlled monitoring of environmental change, to initiatives by researchers and development agents or governments to consult communities about the results of interventions, to the participation of field workers and researchers in evaluation – as opposed to external evaluations by funding agencies with little focus on community involvement. Bottom–up or participatory approaches to developing indicators provide opportunities for community empowerment not otherwise provided by top–down approaches (Fraser *et al.* 2005); they ensure that the indicators are relevant to local situations; they measure what is important to the community; and they are adapted over time as community circumstances change (Freebairn and King 2003; Pretty 1995; Carruthers and Tinning 2003). Involvement of communities helps to build community capacity to address future problems and leads to local action to deal with current problems. These participatory processes, however, have implications for methodology issues; for example, Carruthers and Tinning (2003) and Reed *et al.* (2005) suggest that the methods used to collect, analyse, and interpret data must be easy enough for active participation by local communities.

Reed *et al.* (2005) summarise some methodological frameworks for developing and applying sustainability indicators at local scales. These include the soft systems analysis, which builds on systems thinking and experiential learning to develop indicators as part of participatory learning processes (Checkland 1981); sustainable livelihood analysis, which develops indicators to monitor changes in the natural, physical, human, social, and financial capital (Scoones 1998); Classification Hierarchy Framework, which identifies indicators by incrementally increasing the resolution of the system (Bellows 1995); and The Natural Step, which develops indicators to represent four conditions for a sustainable society (TNS 2004).

Despite being useful for engaging local stakeholders, these participatory processes do have shortcomings. Community indicators can be very specific to communities – based on their experiences and on the local context. This makes it difficult for the wider extrapolation and comparison of indicators across communities and across landscapes. Because of some of these shortcomings, expert-led or external-led approaches argue the need for generic indicators. It is clear that neither of these two approaches is adequate on its own, and there is a need for a hybrid methodology that borrows from these two approaches. In response to Bossel's (2001) framework of a systems-based approach, Reed *et al.* (2005) add to this debate by asking to what extent the formal top–down frameworks can integrate the desire to involve local people, and what compromises have to be made to bridge the two approaches.

While most of the discussion on community indicators and their use has been in the context of environmental sustainability (Reed *et al.* 2005, 2006; Bossel 2001; Fraser *et al.* 2005), we use the same concept of community indicators within a sustainable livelihoods framework. The concept and framework were applied in a community project within a multi-country initiative called Enabling Rural Innovation (ERI), aimed at improving the well-being of communities by strengthening their capacity to self-organise and develop sustainable community agro-enterprises, and to use improved technologies to sustain the enterprises and improve food security. In this project, we integrated a PM&E process into the project, with communities developing a vision for the project, indicators to monitor progress, data-collection tools, and being involved in the data collection, analysis, and interpretation. This article does not give the results of the community process, but rather describes a process of using an aggregation of the community indicators to allow for their comparison across six communities in Malawi, and to apply some basic statistical methods to compare the community perception of the achievement of the indicators across communities and gender. We give a brief description of the ERI initiative, in order to show the context within which PM&E was applied. We then discuss a framework for the PM&E process, analyse the indicators from a general perspective, and finally discuss the differences by gender of the perceptions of the achievements.

The Enabling Rural Innovation initiative

The Enabling Rural Innovation (ERI) initiative is a research-for-development framework which uses participatory research approaches to strengthen the capacity of research and development (R&D) partners and rural communities to access and generate technical and market information for improving farmers' decision making. The aim is to create an entrepreneurial culture in rural communities, where farmers 'produce what they can market, rather than trying to market what they produce', and encourages them to invest in their natural resources rather than depleting them for short-term market gain (Best 2002; Ferris *et al.* 2006).

At community level, emphasis is on strengthening the capacity of smallholder farmers to link effectively to markets, develop profitable agro-enterprises, and test and use technologies to sustain their enterprises and improve food security. The approach is holistic and includes capacity strengthening in the areas of gender, HIV, business skills, and leadership skills. Rather than use an external approach to monitoring and evaluation, PM&E is integrated into the process by building the skills of communities to develop their vision and objectives, and indicators to monitor these. The key steps in implementing the ERI process are shown in Figure 1.

A PM&E process is built in to allow for regular learning and reflection loops between communities and partners, to ensure that lessons are documented and adjustments are made in a timely

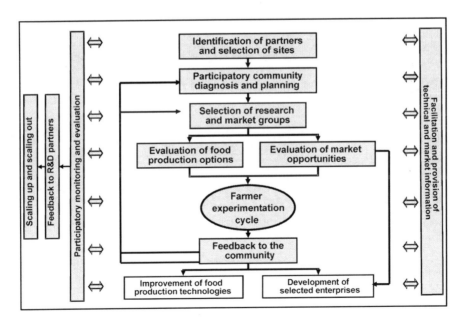

Figure 1: Key steps in Enabling Rural Innovation

manner, providing critical feedback which forms the basis for scaling up the process into agricultural R&D organisations. The objectives of community-based PM&E are to strengthen the collective capacity of rural communities to define their desired outcomes, and their indicators of change, tracking changes and making necessary adjustments, sharing and learning together as they develop their agro-enterprises and experiment with improved technologies.

Community monitoring and evaluation within ERI

Figure 2 provides the general framework that we have used for community PM&E. This framework was used to develop indicators across six communities working within the ERI initiative in Malawi.

While these indicators are community-specific activities, there was a need to compare indicators across all the communities to get a general indication of the extent to which the communities felt that they had achieved their vision and objectives. The indicators from the six communities were aggregated, and a comparison was made across communities.

The short- and long-term results and their indicators were aggregated for all six communities. The results were grouped into three types: livelihood, human capital and empowerment, and social capital (Table 1).

Using cluster random sampling, a total of 86 households was interviewed from the six communities. Each community was taken as a cluster, and a semi-structured questionnaire was used

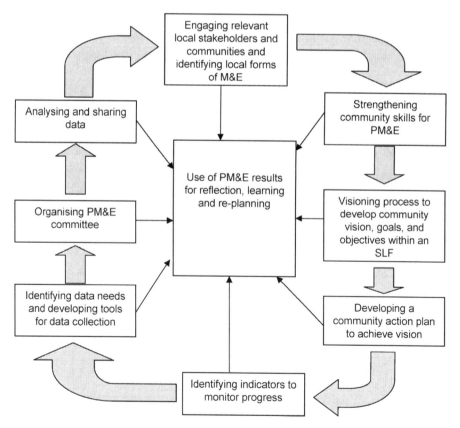

Figure 2: Methodology framework for community participatory monitoring and evaluation
SLF = sustainable livelihoods framework.

Table 1: Aggregated results and indicators from the six communities in Malawi

Types of indicator	Result	Indicators
Livelihood	Food security	• Granaries with maize throughout the year • Cases of child malnutrition in the village
	Increased income	• Better clothing and shoes for families • Ownership of assets (bicycles, carts, mats) • Availability and use of fertilisers
	Increased access to markets for produce	• Obtaining better prices for produce • Ability to negotiate for better prices
Human capital and empowerment	Empowerment of women	• Women speaking up during meetings • Women taking leadership roles in the group and community • Women have the capacity to buy clothes without requesting permission from husbands • Women going to market to sell produce
	Improved gender relations	• Men and women helping each other out • Men and women making decisions in the household together
	Increased knowledge in experimentation	• Increased production of crops such as beans, groundnuts • Diversification into new crops • Ability to solve own agricultural problems through experimentation
Social capital	Group development	• Participation of members in meetings • Members fulfilling their roles in the club • Strength and cohesion of the club
	Improved decision making and confidence	• Capacity to approach extension worker • Self-reliance in finding seeds and other inputs, as well as linking with other organisations • Awareness and behaviour change in relation to HIV and AIDS

to collect data. The questionnaire used a Likert scale with a 5-point scale for individuals to assess the extent to which the project had achieved the expected results. Summaries of the assessments and mean scores were analysed across sites and for both men and women. Next, we discuss the achievement of three outcome-level results (improved food security, increased incomes, and women's empowerment) and consider how perceptions of the changes varied across communities and by gender.

Community perception of changes in indicators

Food security

Two of the key indicators for food security were the existence of granaries with food throughout the year, and a reduction in cases of child malnutrition. A majority of respondents perceived that food insecurity and cases of malnutrition were on the decline (59.3 per cent and 61.6 per cent, respectively). There was a relationship between these two indicators: in sites where more of the households felt that they kept sufficient food throughout the year (from harvest to harvest), the prevalence of child malnutrition was also perceived to be lower (see Figure 3).

Incomes

Many households reported having managed to set up enterprises from which they had obtained income. This was especially true for farmers who were engaged in livestock-based enterprises such as pigs and goats. In addition, other households also stated that they sold groundnuts, beans, and milk from dairy animals. All communities stated that they had bought some assets from the sales of their produce through market linkages established during the project (Figure 4). The majority of respondents (57 per cent) stated that they had better clothing for their families. In addition to buying clothes, many households also purchased assets such as bicycles and kitchen utensils, and fertilisers.

Women's empowerment and gender relations

Four indicators were used to determine farmer perception of changes in women's empowerment and gender relations. While all four indicators were perceived to have changed, those of 'women speaking out during meetings' and 'women taking leadership positions' were perceived to have improved more across all the communities than the other two indicators. Compared with the other three indicators, the ability of women to purchase clothes without asking for permission from their husbands was perceived to have remained the same by more households (Figure 5). This would imply that in some of the households, despite women speaking out and taking on some of the community leadership positions, they were not able to influence the control and use of incomes. There were, however, some changes in the control of income at household level, especially where households were engaged in non-livestock enterprises such as beans and groundnuts, which were still to some extent controlled by women.

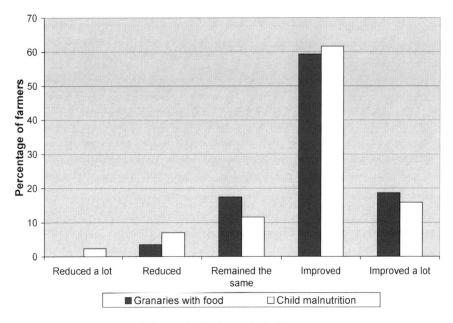

Figure 3: Farmer perceptions of changes in food-security indicators

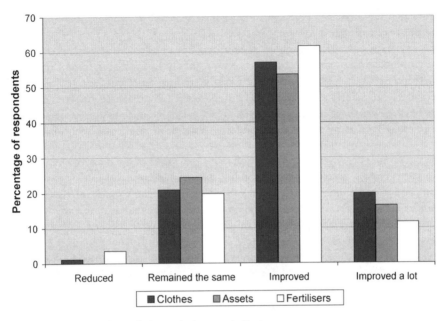

Figure 4: Farmer perceptions of change in income indicators

Gender differences in perceptions of change

There were differences in the perceptions of male- and female-headed households of changes in all three outcomes. All the income indicators were rated higher by men than by women, with men rating 'better clothing' significantly higher than women. The higher rating of income indicators by men could be a reflection of the fact that men still control most of the income from the

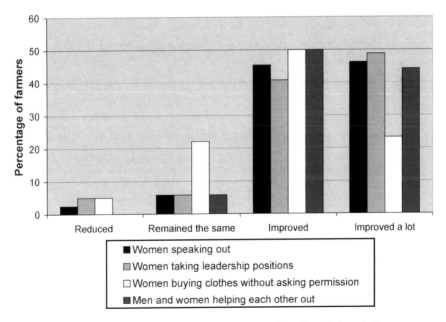

Figure 5: Farmer perceptions of change in empowerment and gender-relations indicators

enterprises and therefore perceive this to have increased, while women (who may not control the income) may feel that income has not improved.

Empirical studies of intra-household gender dynamics elsewhere in Africa have shown that when a crop enters the market economy, men are likely to take over from women, and that women therefore do not benefit from market-oriented production (Quisumbing *et al.* 1998; Kaaria and Ashby 2001; Cornwall 2003). It is argued that market-oriented production is likely to result in increased income controlled by men, despite the fact that such market-oriented production may take away women's labour from food-crop production, especially if the cash crops are different from food crops. There is also a risk that market-oriented production may result in the capture of the benefits by the rich to the detriment of the poor, or create a privileged group of farmers with access to new technologies and markets.

A comparison of the mean scores was made between men and women for all the interviewed households (Table 2). For food-security indicators, the results were mixed, with women scoring significantly higher for granaries with food, while men scored higher for decline in child malnutrition. For the income indicators, women scored higher for better clothing, while there were no significant differences in the scores by men and women for the other two indicators. We separated data for male-headed households and female-headed households, and made an analysis

Table 2: Comparison of mean scores of indicators by gender

Outcome	Indicator	Comparison between all men and women respondents		Comparison between women in female-headed households and women in male-headed households		Comparison between men and women in male-headed households	
		T value	P value	T value	P value	T value	P value
Food security	Granaries with food	2.638	0.100*	−0.303	0.764	2.643	0.010***
	Decline in child malnutrition	−0.442	0.052*	2.388	0.023**	1.301	0.198
Incomes	Better clothing	1.964	0.053*	1.729	0.094*	2.155	0.350
	Accumulation of assets	1.479	0.143	1.729	0.094*	1.797	0.077*
	Use of fertiliser	1.004	0.319	1.826	0.078*	1.823	0.073*
Women empowerment and gender relations	Women speaking out	−0.031	0.975	−0.718	0.479	−0.439	0.662
	Women taking leadership positions	−1.324	0.189	0.178	0.860	−1.512	0.135
	Women buying clothes without asking	−0.198	0.844	1.306	0.201	0.794	0.430
	Men and women helping in household	0.475	0.634	1.308	0.307	0.496	0.621
	Women participating in household decision making	0.924	0.358	0.798	0.431	1.077	0.285

*p < 0.1; **p < 0.05; ***p < 0.01.

of differences in scores between men and women in the male-headed households. Women's scores for granaries with food, accumulation of assets, and use of fertilisers were significantly higher than those of the men. There were also differences between women in female-headed households and women in male-headed households. For all income indicators – that is, better clothing, accumulation of assets, and fertiliser use – women in male-headed households scored higher than women in female-headed households, implying that the increases in income may have been felt more or may have been actually higher in male- than in female-headed households. There were no significant differences in the income.

Conclusions

Community indicators are an effective mechanism for involving communities in PM&E, for ensuring that indicators are locally relevant and measure change according to community perspectives. While we have attempted to show some simple methods through which community indicators (most of which are highly context-specific) can be aggregated and measured across communities implementing similar projects, the need for complementing these with more quantitative methods to support the community perceptions cannot be over-emphasised. Community indicators also need to be grounded in more theoretical frameworks, so that they can be compared with other generic indicators and so that community progress is not assessed from a narrow perspective. As with most changes, the cause-and-effect relationships that determine change are usually impractical if not impossible to define, and the low use of statistical methods in community PM&E does not allow for these to be quantified. Community PM&E therefore needs to be accompanied with the use of participatory tools and methods that allow for inferences and causal relationships to be inferred between the activities and the indicators of change.

References

Bellows, B. C. (1995) 'Principles and Practices for Implementing Participatory and Intersectoral Assessments of Indicators of Sustainability: Outputs from the Workshop Sessions SANREM CRSP Conference on Indicators of Sustainability', Sustainable Agriculture and Natural Resource Management Collaborative Research Support Program Research Report 1/95: 243–68.

Best, R. (2002) 'Farmer participation in market research to identify income-generating opportunities', CIAT Africa Highlights, Kampala: International Centre for Tropical agriculture, available at http://www.ciat.cgiar.org/africa (retrieved 2 October 2007).

Bossel, H. (2001) 'Assessing viability and sustainability: a systems-based approach for deriving comprehensive indicator sets', Conservation Ecology 5(2):12 [online], published at http://www.ecologyandsociety.org/vol5/iss2/art12/.

Carruthers, G. and G. Tinning (2003) 'Where, and how, do monitoring and sustainability indicators fit into environmental management systems?' Australian Journal of Experimental Agriculture 43 (3): 307–23.

Checkland, P. (1981) Systems Thinking, Systems Practice, Chichester: John Wiley.

Cornwall, A. (2003) 'Whose voices? Whose choices? Reflections on gender and participatory development', World Development 31 (8):1325–42.

Ferris, S., E. Kaganzi, R. Best, C. Ostertag, M. Lundy, and T. Wandschneider (2006) A Market Facilitator's Guide to Participatory Agroenterprise Development, Cali, Colombia: Centro Internacional de Agricultura Tropical (CIAT).

Fraser, D. G., A. J. Dougill, W. E. Mabee, M. Reed, and P. McAlpine (2005) 'Bottom up and top down: analysis of participatory processes for sustainability indicator identification as a pathway to community empowerment and sustainable environmental management', Journal of Environmental Management 78 (2): 114–27.

Freebairn, D. M. and C. A. King (2003) 'Reflections on collectively working toward sustainability: indicators for indicators!' Australian Journal of Experimental Agriculture 43 (3): 223–38.

Germann, D. and E. Gohl (1996) *Group-based Impact Monitoring*, Participatory Impact Monitoring Booklet 1, Deutsche Gesellschaft für Technische Zusammenarbeit (GTZ).

Guijt, I. and J. Gaventa (1998) 'Participatory Monitoring and Evaluation: Learning from Change', IDS Policy Briefing 12, Brighton: Institute of Development Studies.

Kaaria, S. and J. Ashby (2001) 'An Approach to Technological Innovation that Benefits Rural Women: The Resource to Consumption System', PRGA Working Document No. 13, Cali, Colombia: PRGA Program.

McAllister, K. (1999) *Understanding Participation: Monitoring and Evaluating Process, Outputs and Outcomes*, Ottawa: IDRC.

Pretty, J. N. (1995) 'Participatory learning for sustainable agriculture', *World Development* 23 (8): 1247–63.

Probst, K. (2002) 'Participatory Monitoring and Evaluation: A Promising Concept in Participatory Research? Lessons from Two Case Studies in Honduras', PhD dissertation, Stuttgart, Germany: University of Hohenheim.

Quisumbing, A., L. R. Brown, L. Haddad, and M. Meinzen-Dick (1998) 'Gender issues for food security in developing countries: implications for project design and implementation', *Canadian Journal of Development Studies* 19: 185–207.

Reed, M. S., E. D. G. Fraser, S. Morse, and A. J. Dougill (2005) 'Integrating methods for developing sustainability indicators that can facilitate learning and action', *Ecology and Society* 10(1): r3, available at http://www.ecologyandsociety.org/vol10/iss1/resp3/ (retrieved 22 September 2007).

Reed, M. S., E. D. G. Fraser, and A. J. Dougill (2006) 'An adaptive learning process for developing and applying sustainability indicators with local communities', *Ecological Economics* 59 (4): 406–18.

Scoones, I. (1998) 'Sustainable Rural Livelihoods: A Framework for Analysis', IDS Working Paper 72, Brighton: Institute of Development Studies.

The Natural Step (TNS) (2004) 'The Natural Step', available at http://www.naturalstep.org/ (retrieved 14 September 2007).

Participatory technology development in agricultural mechanisation in Nepal: how it happened and lessons learned

Chanda Gurung Goodrich, Scott Justice, Stephen Biggs, and Ganesh Sah

International Wheat and Maize Improvement Center (CIMMYT) projects on new resource-conservation technologies (RCTs) in the Indo-Gangetic Plains of Nepal aimed to strengthen equity of access, poverty reduction, and gender orientation in current rural mechanisation processes – more specifically, to promote machine-based resource conservation and drudgery-reduction technologies among smallholder farmers. These projects, together with other projects and other actors, gave rise to an informal 'coalition' project, which used participatory technology development (PTD) approaches, where farmers, engineers, scientists, and other partners worked towards equitable access to new RCTs. This experience showed that PTD projects need to be flexible, making use of learning and change approaches. Once successful adoption is occurring, then what? Such projects need to ensure that everyone is benefiting in terms of social inclusion and equity; this might necessitate new unforeseen work.

Background

Two-wheel tractors (2WTs), also known as power tillers or walking tractors, have been spreading in Nepal since they were introduced in the mid-1970s. Their use was mainly confined to the Kathmandu and Pokhara valleys, where they were used for transport activities and in agriculture during times of peak demand for tillage and services. The tractors were highly remunerative and paid off the capital cost in a short period of time. However, in 1992 there was a ban on new registrations by the local municipalities, due to concerns about traffic congestion; essentially all imports of 2WTs were stopped.

From 1991 to 1996, a USAID-funded project led by the International Maize and Wheat Improvement Center (CIMMYT) tested and promoted the use of new resource-conservation agricultural technologies (RCTs) and minimum-tillage practices that bring improved soil health and reduce production costs in the rice–wheat cropping systems on the *terai* (plains area). The basic power source for most of the technologies was the 2WT. CIMMYT and its partners in the Nepal Agricultural Research Council (NARC) used a farming systems research and extension (FSR&E) approach on agricultural research stations and farmers' fields. It was expected that simply taking developed and 'proven' machinery to the fields of smallholder farmers for demonstration purposes would create demand. However, by 1996 farmers were not adopting the technology. They were not convinced of the usefulness and viability of the machinery being demonstrated. After a review of the work in 1996, one of the reasons identified for non-adoption was that the farmers were allowed only limited participation in the experiments and farmer field demonstrations (Justice 2000). It was proposed that under a new project funded by the UK government's Department for International Development (DFID) the programme would begin to use participatory technology development (PTD) methodologies. At first, NARC was reluctant to participate in the proposed participatory programme, because engineers and scientists were risking government-financed experiments by treading on unfamiliar theoretical and methodological grounds (namely, PTD). Moreover, even in CIMMYT, discourses were still dominated by a linear transfer-of-technology (TOT) culture in research and development. In Nepal this meant that R&D was centralised, top–down, and worked in isolation; considered itself all-knowing; had static programme designs and fixed packages; viewed unexpected outcomes as mistakes and failures; belittled, criticised, and formally punished 'transgressors'; and interpreted as poor performance of scientific duty the admission that farmers, local agricultural extension district officers, or machinery owners with no formal education might have answers to some problems, or point towards priority issues for formal engineering research.

Researcher managers were also wary of giving up operational control of state-owned agricultural machinery to farmer groups. Slowly, however, NARC engineers and scientists began to allow the PTD-based research to progress, as they perceived that – by working with an outside agency like CIMMYT in a special project mode – they had a layer of protection from the normal TOT culture of the NARC and some international agricultural research systems. Still, overcoming personal and institutional issues and uncertainties was no minor feat, as those involved in the informal coalition project were entering areas that were uncharted for all concerned.

Broader rural mechanisation policy considerations

Past failures to investigate, understand, and publicise perceived agricultural engineering successes and failures (both national and international) from the 1960s to the 1980s resulted in the withdrawal of government and donor funding for agricultural mechanisation research, development, and extension (RD&E) (Justice 2000; Starkey 1989). In the 1990s, with the spread of structural adjustment policies there was also a belief that 'free' markets would induce 'appropriate' mechanisation technologies.

In this RD&E vacuum, the uninformed private industries and importers selected agriculture machinery according to what they supposed they could sell easily and quickly, and this mostly resulted in a lack of appropriately sized and affordable agricultural machinery for smallholder farmers, and *de facto* the promotion of a Western-style agricultural mechanisation process in many parts of South Asia. Farmers and manufacturers missed the chance to try out and evaluate new mechanisation technologies, due to continued lack of interest in and funding of mechanisation research.

The PTD approach required additional skills in project management and coalition building among stakeholders that are not normally associated with international agricultural research organisations such as CIMMYT.

Due to the lack of interest in agricultural mechanisation research since the early 1980s, international public-sector agricultural engineering as a discipline was often not included in FSR&E and PTD initiatives. Consequently, agricultural engineers failed to acquire and apply PTD skills and methodologies in their R&D programmes (Justice 2000).

Researchers and engineers failed to recognise that poor farmers and rural labourers could benefit from rural mechanisation by acting as tractor service providers, but only if resource-conservation technologies like the two-wheel tractors were made accessible to them (Biggs *et al.* 1993).

The new coalition research approach

Faced with challenges such as lack of imported machinery and the intensification of the Maoist conflict, by 1999 the PTD project had begun to transform itself into a coalition project, with many linkages to other projects, government programmes, NGOs, and the private sector. By working with different partners, those involved have changed their behaviour in many ways, such as the following.

The farming systems TOT approach became a more interactive PTD approach, with significant results: farmers began to experiment with the machinery alongside scientists and engineers, and trained farmers were used as resource persons for new groups; private-sector agro-machinery importers and potential importers were invited to PTD sites and technical training events to interact with farmers to learn of the potential demand from farmer groups; the banking sector was invited to see the economic viability of the technology so that they might develop 'new financial products' to provide credit to target farmer groups.

The coalition project began to focus on poverty-reduction and gender-equity issues. It started to work with partners who were also committed to poverty-reduction and gender-equity goals, but who were not formally part of the designated projects. There was a recognition that projects are rarely standalone activities with clear ways of meaningfully defining their own inputs, outputs, monitoring, and evaluation procedures.

There have been significant methodological innovations, including the following.

- New groups identified and included the poorest farmers in the village. Of the five-member formal management committee, at least two members were women. In addition, each group nominated two village motivators, one of whom had to be a woman. Two new farmers' groups had to have 60 per cent or more women as members. Promotion and evaluation of the group management approach increased the number of farmers gaining experience with the 2WT and rotovator.
- With help from the Consultative Group on International Agricultural Research (CGIAR) Systemwide Program on Participatory Research and Gender Analysis for Technology Development and Institutional Innovation (PRGA Program), specialised training for scientists and practitioners was given (at various times) on gender analysis and participatory research and development concepts, principles, and management.
- Staff of another project helped to organise six-monthly review and planning meetings, where the scientists and practitioners share experiences and plan for the next six months. This includes the use of management and monitoring tools, such as Actor Linkage Maps, Linkage Matrix, Time Lines, and Learning and Response Tables.
- Annual national meetings of scientists and practitioners, NGOs, government agencies, farmers, business persons, machine operators and mechanics, machinery owners, bankers, donors, and others were organised. One particularly helpful workshop was sponsored by Nepal Society of Agricultural Engineers. These national meetings helped to move policy

discourses forward from the out-of-date, simplistic, but ever-present arguments in favour of or against rural mechanisation towards an analysis of current problems and possible solutions, regarding pro-poor rural mechanisation policies and activities.

- Studies of the urban-based 2WTs innovation systems in Kathmandu and Pokhara valleys were undertaken to see what could be learned from the viable systems that were already spreading. An informal understanding was developed with a parallel project which concentrated on evaluating the gender outcomes of these types of mechanical intervention.

Key outcomes from the coalition project

Equitable access

At the start of the project, it was thought that only members of the groups formed (20–30 members from different households) would use the 2WTs and accessories, yet as the project progressed we found that the members were also renting out the technologies to non-members. A survey conducted at the end of one winter cropping season showed that almost as many, if not more, non-members also used the 2WTs. Depending on their financial status, users of 2WTs were not always expected to pay immediately; rather the tractors could be hired on credit and paid after the harvest had been sold – a benefit not normally available from the providers of four-wheeled tractors operating outside the community. This financial arrangement is a traditional system within the village, and in this case high interest charges were not imposed.

Gender relations

Men were able to help women with household chores and parenting because use of the SWTs meant that they had to spend less time and labour in ploughing. Women were able to assist men in farming because they had to prepare and take food to them or to hired ploughmen in the field less often. Women liked the 2WTs because they did not need to dig out the corners of the field by hand, as they would with land preparation using larger four-wheel tractors. They also said that they were keen to learn how to drive the 2WT (Gurung and Justice 2007; Page 2007). Indeed, in response to requests from women within the project in the first year, it was decided to form a new 'all-women group' in Bahuwari village in Bara District, where women were trained to manage and operate the 2WTs. Problems were predictably plentiful, but after further investigations there was evidence that there was enough social and cultural room to manoeuvre that would allow women to own, manage, and even operate large agricultural machinery (Gurung and Justice 2007).

Increased production

Farmers reported an increase of 5–15 per cent in wheat harvest, which increased some poor households' food self-sufficiency from three months to six months.

Multi-cropping

We found evidence that with the use of the 2WTs and new RCT attachments, cropping intensity is increasing. Some farmers planted maize after harvesting rice and before planting mustard. Others added dry/hot (spring) season crops after wheat in fields that in the past had remained fallow. It seems that the time saved in land preparation and other activities afforded by the use of 2WT and attachments has freed up space within the system to grow additional crops.

Reduced production costs

Farmers reported that production (tillage and planting) costs were reduced by between a third and a half.

Rural transportation

We observed the 2WT with trailer attachment providing much-needed rural transportation facilities such as taking children to school, hauling clay from fields for repairing homes, carrying construction materials from main bazaars to villages for new 'cement' home construction, transporting of manure, fertiliser, and seed to the fields, and transporting straw, grains, and other produce from fields to homes and markets.

Importers, buyers, and entrepreneurs

Across Nepal, as many as 12 importers are now selling 2WTs, and some ten local workshops are selling locally manufactured trailer attachments for rural transport. As of 2008, nearly 7000 2WTs have been sold. Informal assessments have estimated that many buyers of the 2WT are smallholder farmers cultivating less than two hectares, for most of whom it is the first purchase of such a machine, and more than 80 per cent purchase the machinery with the intention of renting out their services. More than 50,000 farm families are hiring the tillage and transport services of these 2WT entrepreneurs. These farm families save more than 33 per cent on crop-establishment costs (compared with the cost of hiring bullocks or four-wheel tractors) and cultivate more than 100,000 hectares per year.

Advocacy

A great deal of effort has gone into changing the Nepalese government's *laissez-faire* policy towards agricultural mechanisation. The debate has now moved on, and discussion of appropriate agricultural mechanisation issues is now more informed among senior policy makers, bureaucrats, aid agencies, and international consultants.

Group management of technology

The group-management approach experienced great difficulties, and the hope of promoting group ownership was never realised. However, because of the high costs of the machinery for poor households, providing such a valuable set of machinery to a single individual would not be fair and could be construed as favouritism or discrimination on the part of the researchers. The group approach was successful in demonstrating these technologies to a wider audience, which would have been impossible under single-owner management.

Linkages

One of the outcomes of the coalition project was that it explored and demonstrated in a practical way the potential benefits of collaboration between different ministries, agencies, and others. The coalition project demonstrated that institutional changes in many areas that benefit poorer and marginalised rural people can be encouraged when attention is given to these issues in policy and project practice.

Lessons learned

Development requires flexibility

A major rationale for using 2WTs in this area was to promote minimum-tillage operations. Researchers were at first afraid that the 2WTs might be used solely for transport. Through the survey of operators of the urban-based 2WTs in the Kathmandu valley and interactions with the farmers and entrepreneurs, the researchers came to realise that non-agricultural uses, such as transportation/haulage in the agricultural off-season, were critical for the viability, profitability, and loan repayment of 2WTs. When trailers and the 'full tillage' rotovators were added to 2WT demonstrations, farmers and entrepreneurs immediately began buying the 2WTs.

Project partners remarked that openness to the overlapping but complementary nature of projects is useful in adopting a flexible 'learning and change' approach. The project is also monitoring and learning from rural mechanisation innovations that are taking place elsewhere in the region, especially in Bangladesh and India.

Social inclusion and equity

The management of this coalition project was often not easy. When new partners joined, they sometimes created 'problems' for other partners – for example, when an evaluation project started to investigate gender and social inclusion issues more thoroughly and information that was unwelcome to some other partners was revealed. However, the fact that the partners remained in the coalition speaks well for the overall ethos of the alliance on issues which at many levels are highly contentious because they relate to social power structures and changing institutions.

The problem of increasing the access of the poor to credit

One of the biggest obstacles to improving poor people's access to these RCTs is the access to credit to buy a 2WT. Despite the proven benefits and rapid spread of 2WTs and the efforts of various project staff and linked projects in the National Planning Commission, the Ministry of Agriculture, and the local private and public banks, there is little evidence that new institutional arrangements are emerging to enable poorer rural people to gain access to fair and reliable credit to buy 2WTs.

The attribution problem

The coalition project drew attention to the ever-present problem experienced by staff of aid agencies, bureaucracies, and scientific institutions with managerial assessment procedures, especially of the quantitative types, which make claims to attribute cause and effect in certain inputs and outcomes. In the coalition project discussed, it would be very hard for example to separate out the effects of the project inputs from particular projects, and the effects of the long history of the spread of 2WTs in Nepal. To some extent, this coalition project is riding a wave as regards the spread of 2WTs, which has a long history. However, the project is to some extent helping to steer this type of mechanisation in a pro-poor and equitable direction. It is unlikely that the people involved in this 'coalition' project would have come together and done useful things if each of the actors had participated only in order to be attributed 'a share' of the 'outcomes'.

References

Biggs, S. D., A. P. Kelly, and G. Balasuriya (1993) 'Rural Entrepreneurs, Two-Wheel Tractors and Markets and Services: A Case Study from Sri Lanka', Discussion Paper #242, Norwich: School of Development Studies, University of East Anglia.

Gurung, C. and S. Justice (2007) 'Empowerment of Women through External (Project) Intervention: Case Study of Women Two-wheel Tractor Operators', available at http://www.wocan.org/document_pdfs/IDP46a767c38b14a.doc (retrieved 7 February 2008).

Justice, S. E. (2000) 'Farmer Participatory Research in Agricultural Machinery R&D: A Case Study from Nepal', A Practicum Report submitted in partial fulfilment of the requirements for the degree of Master of Arts, University of Kentucky.

Page, Samantha (2007) 'Nepal's Final Report to DFID under Reaping the Benefits: Assessing the Impact and Facilitating the Uptake of Resource Conserving Technologies in the Rice-Wheat Systems of the Indo-Gangetic Plain', available at http://www.research4development.info/PDF/Outputs/Misc_Crop/Nepalfinalreport.pdf (retrieved 7 February 2008).

Starkey, P. (1989) 'Perfected yet Rejected: Animal-Drawn Wheeled Tool Carriers – A Cautionary Tale of Development', Eschborn: GTZ, and Braunschweig: Vieweg.

Gender equity and social capital in smallholder farmer groups in central Mozambique

Elisabeth Gotschi, Jemimah Njuki, and Robert Delve

This case study from Búzi district, Mozambique investigated whether gender equality, in terms of male and female participation in groups, leads to gender equity in sharing of benefits from the social capital created through the group. Exploring the complex connection between gender, groups, and social capital, we found that gender equity is not necessarily achieved by guaranteeing men and women equal rights through established by-laws, or dealing with groups as a collective entity. While there were no significant differences in the investment patterns of men and women in terms of participation in group activities and contribution of communal work, access to leadership positions and benefits from social capital were unequally distributed. Compared with men, women further found it difficult to transform social relations into improved access to information, access to markets, or help in case of need.

Introduction

The benefits of collective action in agricultural activities and farmer groups (for example, bargaining, improved access to technology, and reduction of transaction costs) have been well recognised in the literature (see, for example, Pretty and Ward 2001; Westermann *et al.* 2005). However, while gender has become increasingly important in the development debate, gender relations within groups have not been adequately addressed. Inter-relationships between gender, collective action, and social capital are just beginning to be explored (Adkins 2005; Agarwal 2000; Molyneux 2002: 177; Westermann *et al.* 2005).

The debate about social capital

Social capital is one asset that individuals produce through interacting with each other, creating relationships of trust and common understanding. Unlike physical capital, social capital is a 'social' resource, in the sense that it is only accessible to actors through interaction with others (Grootaert *et al.* 2004). High levels of social capital facilitate the development of

shared norms within social groups and networks (Grootaert *et al.* 2004; Uphoff and Wijayaratna 2000), and can improve the efficiency of society by facilitating co-ordinated actions (Putnam 1993: 167). Owing to its comprehensive character, scholars agree 'that social capital is not a single entity, but is rather multi-dimensional in nature' (Grootaert *et al.* 2004: 3). Different types of social capital have been distinguished: for example, structural and cognitive social capital. Structural social capital refers to social networks, roles, rules, and interaction patterns that are relatively objective and visible; it includes institutionalised forms such as group membership, as well as informal networks and loose contacts. The norms, trust, attitudes, and beliefs which are based on subjective, mental processes that are shared within a group or in society constitute cognitive social capital (Uphoff and Wijayaratna 2000: 1876).[1]

Male and female social networks differ from each other as a result of different gender roles and cultural norms, such as responsibilities within the household, extra-household activities, and division of labour (Agarwal 2000; Westermann *et al.* 2005). These 'different kinds of and qualities of social capital' (Westermann *et al.* 2005: 1785) for men and women are rarely discussed in the literature. While most of the social-capital literature assumes the family as a primary source of social capital (see, for example, Bourdieu 1987), it has been criticised by Adkins (2005), who suggests that the implicit assumption of a family model based on a heterosexual couple resting upon the traditional division of labour and domestic femininity reinforces female subordination and ignores the complexity of the realities of many women all over the world.

In the analysis of gender and social capital, it remains unclear how male and female members differ in terms of investing in groups, especially their willingness to contribute money and labour, participation in group meetings, length of membership of groups, or helping other people. The differences in benefits from group membership experienced by men and women, especially in terms of increasing personal networks, creating supportive social networks, access to information and services, or gaining access to reputable leadership positions, are not documented. Emphasis by NGOs that are trying to achieve gender equity has been to encourage men and women to participate in groups and to achieve gender parity in the numbers of men and women in the groups. In most groups, established by-laws do not distinguish between male and female leaders, rather (in theory) the constitution guarantees equal rights. This article analyses the gendered differences in investments in groups and benefits from social capital, and asks whether women can increase their benefits by entering leadership positions. We also look at the role of women's participation in increasing social capital within groups.

Forms of social capital in Mozambique

Compared with other countries in Southern Africa, social networks in Mozambique are unique as a result of various factors. Due to low population density, settlement patterns are scattered and farmers often live long distances from each other. Consequently, people predominantly organise their social life around their kin. The extended family provides social protection, as government services (health services, child care, and pensions) are rarely in place (Ministério do Plano e Finanças 1998). In addition, social ties within the communities were largely destroyed during the armed civil war (1984–1992), which displaced about 50 per cent of the population.

Perhaps surprisingly, systems of informal co-operation and structure of rural society hav survived the dramatic changes. The main types of informal co-operation between househol include *xitique* (saving and credit), *ajuda mutua* (mutual assistance in daily work), *buscato* or *ganho-ganho* (exchange of labour for money, food, or traditional drinks) (M'

2003). Formal co-operation is comparatively rare, with only 2 per cent of the three million smallholdings being officially organised in groups (Mole 2003: 140). However, with the increased NGO and government activities after the war, growing numbers of non-registered groups are emerging all over the country.

Methodology

The study was carried out in Búzi district of Sofala province in Mozambique. It utilised a variety of tools. A group inventory was carried out of 73 farmer groups, and detailed information on membership was elicited for 20 of the groups with a total membership of 491 farmers. Focus-group discussions were carried out with the 20 groups, and structured interviews were conducted with 160 farmers. The sample of 160 farmers was obtained through a two-stage quota sampling which selected the 20 groups based on quota for geographic location, gender composition, age of group, and other factors, and then eight respondents per group were selected by quota for gender and leadership. The survey instrument was developed from a review of existing measurement tools, adapted to the regional context and research interest (Grootaert *et al.* 2004) to explore insights into the complex inter-relationships among groups, gender, and social capital.

Results and discussion

Equal numbers, equal chances?

Overall, the farmer groups studied in Búzi district were characterised by ethnic homogeneity (96 per cent of respondents belonged to the ethnic group, Ndau) and a gender balance (53 per cent women) in membership figures. However, farmer groups were quite heterogeneous in their socio-economic profile, and most indicators were significantly different between men and women: women were younger (37.8 years vs 46.1 years), had a higher level of illiteracy (31 per cent vs 80 per cent) and had been school-educated for a shorter period of time (1.8 years vs 3.6 years) than men. Women in permanent relations[2] were less likely to become group members, compared with single, divorced, or widowed women. Upon marriage, the construction of female identity includes the woman's subordination, restricted mobility, and dependency on the male household head. Women's relatively low access to financial and human capital can restrict their active participation in groups.

It was reported that typically the husband would become a member, pay the monthly membership fee that most groups impose on their members, and represent the household. In female-headed households, a woman's autonomy increases; but, while such women have the advantage of being members in their own names, female-headed households face greater risks of being poor (UNDP 2001: 46). In contrast, groups perpetuate gender relations and social imbalances that are prevalent in society: women report not being able to talk freely or disagree in front of men; some participate in groups against the will of their husbands, or participate in group meetings outside their village. In focus-group discussions, women indicated that husbands do not like the fact of not knowing where the women are, and therefore women find it hard to leave the household for longer than a couple of hours; their time is also constrained by their multiple tasks and responsibilities (such as household tasks, child care, and farming).

Besides restricting women's personal engagement in the group, traditional gender roles affect women's access to leadership positions (Figure 1). Women are less likely to be president and hence hardly have a chance to represent the group, participate in meetings or seminars, or take final decisions. This is also true for the positions of vice-president and the secretary.

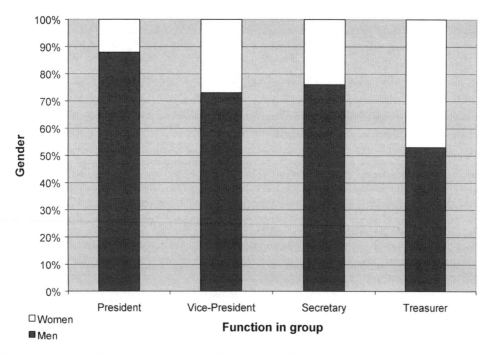

Figure 1: Leadership position disaggregated by gender for 20 farmer groups (n = 491) in Búzi district, Mozambique

Only the position of the treasurer is more often held by women, who are said to be more trust-worthy and less likely to abuse ('eat') money; but even here there is only approximate parity between the sexes (Figure 1). By-laws do not formally define criteria for group leaders; however, it was repeatedly reported that informal criteria were considered for a position; for example, having a social reputation ('being a good person'), trustworthiness ('not eat the money'), ability to read, write, and be able to represent the group. Group members who already occupy positions in the community (such as traditional leaders) are more frequently elected into these positions, prove their ability to deal with authorities, and further improve their social status according to the principle 'them as has, gets' (Putnam 1993: 169).

Gender equity and social capital

We did not find significant differences between men and women regarding their willingness to invest time and money in the achievement of common goals (altruist orientation), participation in group meetings, or contribution to community work (Table 1). Men reported higher trust in other people and a higher degree of helping other people than women, contradicting the common assumption that women in general display more solidarity behaviour (Cornwall 2003; Westermann et al. 2005). A comparison between male group members and female leaders found that these gendered differences are partly overcome when women enter leader-ship positions, although traditional gender roles still limit women's ability to make friends and create relations.

Female members of mixed groups had significantly higher trust in other people, compared with members of female-only groups. Despite equal investments of men and women into

Table 1: Gendered differences in contributing to farmer groups and social capital

Variable	Female (1) vs. male (2)	Female leaders (1) vs. male non-leaders (2)	Women groups (1) vs. women in mixed groups (2)
Altruist orientation†	0.089	−0.021	0.145
Contribution to community work‡	−0.095	0.051	−0.034
Helping other people‡	0.188**	0.076	0.112
Number of friends‡	0.446***	0.392***	0.014
Participation in meetings‡	−0.034	−0.145	−0.033
Trust in other people†	0.401**	0.110	0.352*

Note: A positive value indicates that (2) is better off than (1); and a negative value indicates that (1) is better off than (2) in each comparison.
Significance levels: ***$P < 0.01$; **$P < 0.05$; *$P < 0.1$.
†Gamma; ‡Pearson's R.

groups, men were more successful in benefiting from social capital, measured in terms such as having people who help or provide credit in case of need, number of contacts, and likelihood to access institutions (for example, markets, government, service providers), or information about (for example) technologies and markets. Men were also less likely to report suffering from problems, compared with women. Women in leadership positions could improve their social-capital benefits, attaining to the level of male non-leaders, but it remained more likely that male members, rather than female leaders, would obtain credit (Table 2).

Women in mixed groups were more likely to benefit from social capital (such as access to institutions, sources of help) and report fewer problems, compared with women in female-only groups. Although we suggested above that mixed groups perpetuate female subordination

Table 2: Gendered differences in benefits from structural social capital

Variable	Female (1) vs. male (2)	Female leaders (1) vs. male non-leaders (2)	Women in women groups vs. (1) women in mixed groups (2)
Number of people who give credit†	0.321***	0.341**	0.101
Number of people who help in need†	0.278***	0.221	0.202*
Index§ Source-help-in-need‡	0.162	0.107	0.482***
Index§ Number of contacts‡	0.370***	0.258	0.299*
Index§ Access institutions‡	0.233**	−0.079	0.441**
Index§ Sources of information‡	0.235**	0.232	0.265*
Index§ Number of problems‡	−0.235**	−0.297	−0.365**

Note: A positive value indicates that (2) is better off than (1); and a negative value indicates that (1) is better off than (2) in each comparison.
Significance levels: ***$P < 0.01$; **$P < 0.05$; *$P < 0.1$.
†Pearson's R; ‡Gamma; §Indices are Σ of battery of dichotomous question.

and restrict their participation in leadership positions, the creation of women-only groups addressed only part of the 'gender problem'. Gendered differences are also reflected in distinct social networks of men and women within power relations; such differences determine access to, for example, traders or political institutions. Women in mixed groups found it easier to tap some of the male resources, enter the 'masculine social spaces' (Molyneux 2002: 181), and establish contacts, gain access to information, and obtain help in case of need. In order to achieve gender equity, the challenge ahead is to transform power relations between men and women.

Conclusion

While women in Búzi district form more than 50 per cent of group members, attend group meetings, and invest money, labour, and time in group activities at an equal level to the men, group membership cannot be examined without taking into account the multiple roles of men and women in their households and communities.

Despite formally defined equality in by-laws and gender equity in membership figures, groups in Búzi district perpetuate patriarchal power structures prevalent in society. Women face restricted chances of being elected as group leaders. Traditional subordination under men further restricts their ability to put their issues on the group agenda for discussion, or to capture benefits such as increasing their networks, and accessing information and help in case of need (Cornwall 2003: 1330). Increasing women's participation in groups and leadership positions is a key step in involving women. It does not, however, address power issues and the ability of women to take decisions or put their issues on the group's agenda. Efforts to increase the number of female leaders need to account for women's relatively more complex responsibilities. In addition, having women in committees and leadership positions in groups, although necessary, is not sufficient for opening up space for women (Cornwall 2003). Similarly, it has been demonstrated that the creation of women-only groups avoids addressing the 'gender problem'; by creating a 'gender-free space', the groups fail to fulfil their potential to transform power relations between men and women.

Mixed groups formally provide equal chances for men and women; however, there is a need to challenge traditional gender roles (Kusakabe *et al.* 2001). Gender-sensitive group approaches require that NGOs rethink their strategies and consider the complex relations between men and women, and their respective commitments to the household and community. This not only requires empowerment of female members of farmer groups, but starts from training their own development and extension agents to be aware of gendered differences within groups and to have the capability to deal in a more critical way with collective entities (such as households, communities, and groups). An increased number of members could be used to broaden the scope of group objectives and establish sub-committees for different tasks, allowing a greater number of members to gain experience and skills.

Acknowledgements

The authors are grateful to the farmers in Búzi district for participating in the research, as well as to the project *Promoção Económica de Camponeses* (PROMEC) and the civil-society organisation *União Distrital de Apoio ãos Componeses* (UDAC) for their logistical support in the carrying out of the field work to collect the data reported in this article. Many thanks also to Stefan Vogel for hosting the corresponding author at the Institute of Sustainable Economic Development, University of Natural Resources and Applied Life Sciences, and for making comments on the original version. The study was funded by the government of Austria.

Notes

1. Social capital has been conceptualised differently by various scholars (for example, distinction of bonding, bridging, and linking social capital, or strong ties vs weak ties).
2. The category 'permanent relation' comprises officially registered monogamous or polygamous marriages and 'informal', traditional weddings where *lobolo* (bride price) has been paid, but no official documents received.

References

Adkins, Lisa (2005) 'Social capital: the anatomy of a troubled concept', *Feminist Theory* 6 (2): 195–211.

Agarwal, Bina (2000) 'Conceptualising environmental collective action: why gender matters', *Cambridge Journal of Economics* 24 (3): 283–310.

Bourdieu, Pierre (1987) 'Die feinen Unterschiede. Kritik der gesellschaftlichen Urteilskraft', Frankfurt am Main: Suhrkamp.

Cornwall, Andrea (2003) 'Whose voices? Whose choices? Reflections on gender and participatory development', *World Development* 31 (8): 1325–42.

Grootaert, Christiaan, Deepa Narayan, Veronica Nyhan Jones, and Michael Woolcock (2004) 'Measuring Social Capital. An Integrated Questionnaire', Washington, DC: World Bank.

Kusakabe, Kyoko, Chan Monnyrath, Chea Sopheap, and Theng Chan Chham (2001) 'Social Capital of Women in Micro-vendors in Phnom Penh (Cambodia) Markets: a study of vendors' association', UMP-Asia Occasional Paper No. 53, Pathumthani, UNDP/UNCHS (Habitat)/World Bank, available at http://www.serd.ait.ac.th/ump/OP053.pdf (retrieved 25 September 2006).

Marsh, Robin (2003) *Working With Local Institutions to Support Sustainable Livelihoods*, Rome: Food and Agriculture Organization (FAO).

Ministério do Plano e Finanças (1998) 'Mecanismos de Ajuda Mútua e Redes Informais de Proteção Social. Estudo de Caso das Províncias de Gaza e Nampula e a Cidade de Maputo', Maputo: República de Moçambique.

Mole, Paulo Nicua (2003) 'Land tenure and use', in Instituto Nacional de Estatística (ed.) *Censo Agro-Pecuário 1999–2000. Resultados Temáticos*. Maputo: Instituto Nacional de Estatística.

Molyneux, Maxine (2002) 'Gender and the silences of social capital: lessons from Latin America', *Development and Change* 33 (2): 167–88.

Pretty, Jules and Hugh Ward (2001) 'Social capital and the environment', *World Development* 29 (2): 209–27.

Putnam, Robert (1993) *Making Democracy Work: Civic Traditions in Modern Italy*, Princeton, NJ: Princeton University Press.

United Nations Development Programme (UNDP) (2001) *Mozambique. Gender, Women and Human Development: An Agenda for the Future. National Human Development Report*, Maputo: UNDP.

Uphoff, Norman and C. M. Wijayaratna (2000) 'Demonstrated benefits from social capital: the productivity of farmer organizations in Gal Oya, Sri Lanka', *World Development* 28 (11): 1875–90.

Westermann, Olaf, Jacqueline Ashby, and Jules Pretty (2005) 'Gender and social capital: the importance of gender differences for the maturity and effectiveness of natural resource management groups', *World Development* 33 (11): 1783–99.

Further resources for participatory research and gender analysis

Guy Manners

So much work has been done on participatory research and gender analysis – their implementation, evaluation, and institutionalisation – that it is difficult to recommend a limited set of resources. The context here is 'challenges to operationalising participatory research and gender analysis', so we have sought out resources which shed light on some new practical issues and are based on empirical evidence. Some of the classics in the field have also been included. Readers will find additional resources in and through the bibliographical references of articles included in this issue.

This list was compiled by Guy Manners from annotated submissions by contributors to this special issue.

Publications

Books, articles, and reports

Ashby, Jacqueline Anne
'Methodology for the participation of small farmers in the design of on-farm trials'
Agricultural Administration and Extension 22 (1) (1986): 1–19.
Ashby evaluates three approaches to farmer participation in defining criteria for the design of on-farm fertiliser trials. She describes the methodologies and compares the resultant experimental designs. Increased farmer participation led to significant changes in the design of on-farm trials as a result of insights into how farmers themselves wanted to evaluate fertilisers, and raised basic research questions about improvements in the technology. Ashby concludes that farmer participation in the design of on-farm trials requires fewer resources and less time than diagnostic survey research, yet improves the quality of feedback between scientists and farmers.

Averill, Deborah, Nina Lilja, and Guy Manners (eds.)
Participatory Research and Gender Analysis in Agricultural and Natural Resource Management Research: A Selected Review of the Literature, Version 1
Cali, Colombia: PRGA Program, 2006, ISBN: 978-958-694-089-4, 59 pp., available at http://www.prgaprogram.org/modules.php?op=modload&name=DownloadsPlus&file=index&req=getit&lid=241 (retrieved 17 April 2008)
Includes sections on 'Evaluation of participatory research and gender analysis methods' and 'Use of participatory methods in impact assessment, monitoring and evaluation'.

Barrientos, Stephanie and Catherine Dolan (eds.)
Ethical Sourcing in the Global Food System: Challenges and Opportunities for Fair Trade and the Environment
London: Earthscan, 2006, ISBN: 1 84407 199 5 / 1 84407 189 8, 192 pp.
This book asks many questions about ethical sourcing, which is becoming a major concern in 'Northern' food retailing. Large supermarkets are under pressure to ensure that a fair price is paid to small producers and to maintain good employment conditions within supply chains. But how effective is ethical sourcing in addressing the problems facing producers and workers? Can workers' and producers' rights and participation be improved, given the power and dominance of large supermarkets within the global food chain? What role can civil society and multi-stakeholder initiatives play in ensuring the effectiveness of ethical sourcing? This book explores the challenges and opportunities in ethical food sourcing through analysis and case studies which examine a range of approaches.

Barton, Carol
'Where to for women's movement and the MDGs?'
Gender and Development 13 (1) (2005): 25–35.
Available at: http://www.oxfam.org.uk/what_we_do/resources/downloads/gmd-4.pdf
Barton considers the responses of women's-rights activists and organisations to the Millennium Development Goal (MDG) agenda and processes. She reviews the then current state of play of engagement of women's movements with the MDGs, then focuses on campaigning and advocacy, and grassroots activism movements; women's critiques of the MDGs; and some of the ways in which women are choosing to engage with the MDGs, to advance their own agenda.

Bebbington, Anthony, David Lewis, Simon Batterbury, Elizabeth Olson, and M. Shameem Siddiqi
'Of texts and practices: empowerment and organisational cultures in World Bank-funded rural development programmes'
Journal of Development Studies 43 (4) (2007): 597–621.
The World Bank has become concerned about 'empowerment', following on from its discussions of participation, NGOs, and civil society. Commitments to empowerment enter World Bank texts with relative ease, but their practice within Bank-funded projects is less predictable, and they assume diverse meanings. This paper analyses the 'organisational cultures' of the Bank and many of the organisations that implement its rural development programmes. It presents a framework for analysing organisational cultures in terms of (a) the broader contexts in which organisations (and their staff) are embedded; (b) the day-to-day practices within organisations; (c) the power relations within and among organisations; and (d) the meanings that come to dominate organisational practice. A case study of a development programme in Bangladesh illustrates the ways in which cultural interactions among diverse organisations – the World Bank, government agencies, NGOs, organisations of the poor, social enterprises – translate textual commitments to empowerment into a range of diverse practices.

Bentley, Jeffrey W.
'Fact, fantasies, and failures of farmer participatory research'
Agriculture and Human Values 11 (2&3) (1994):140–50.
Bentley asserts that farmer participatory research (FPR) has probably been of interest more because of dissatisfaction with the Green Revolution and agricultural establishment research than because of a proven ability of scientists and farmers to collaborate together. He notes that there are several barriers between farmers and scientists, especially social distance. He says that

the role of FPR should be critically examined: it may work best in setting research agendas or in the case of researchers who can dedicate themselves full-time to FPR for some time.

Biggs, Stephen and J. Farrington
Agricultural Research and the Rural Poor: A Review of Social Science Analysis
Ottawa: International Development Research Centre, 1991, ISBN: 088936573 3, 140 pp.
The authors review a range of social science and economic methods and theories used in the analysis of agricultural and natural-resources research and development, and consider the relevance of these methods to *ex-post* and *ex-ante* analysis of poverty reduction.

Blackmore, Chris, Ray Ison, and Janice Jiggins (eds.)
Special Issue 'Social Learning: an alternative policy instrument for managing in the context of Europe's water', *Environmental Science and Policy* 10 (6) (2007): 493–586, ISSN: 1462-9011.
The focus of this special issue is policy-oriented research, examining non-coercive natural-resource governance at catchment scales. It presents a critical review of social learning theory and its application in practice. It elaborates methodological approach to and techniques for researching the effects and impacts of processes of systemic change, and for understanding the role of science in such processes. Its country coverage is England, Scotland, France, The Netherlands, and Italy, with reference to US and Australian experience. It offers insight into the management of knowledge and knowledge-generating processes, across a range of scales of interaction, and among diverse organisational and individual actors. It also demonstrates how to research the processes at work and assess their effects. It offers important 'breakthrough' insight about how to move beyond conventional impact assessments, with wide application beyond the cases examined.

Bryman, Alan
Social Research Methods, 2nd rev. ed.
Cambridge, UK: Cambridge University Press, 2004, ISBN: 019926446 5, 608 pp.
This is a textbook, claimed by *Communication Booknotes Quarterly* to provide 'an insightful overview of the characteristics and applications of focus groups'. The book aims to explain and demonstrate the main theories and techniques in social research methods, through real-life examples and student learning aids. The text is accompanied by a comprehensive companion website.

Chambers, Robert
Challenging the Professions: Frontiers for Rural Development
London: ITDG, 1993, ISBN: 185339194 8, 180 pp.
Challenging the Professions provides development workers with an agenda for changing themselves before they attempt to change the rural poor. It builds on Chambers' earlier work, *Rural Development: Putting the Last First*, by further questioning the behavioural norms of professions, disciplines, and bureaucracies involved in rural development. Chambers proposes that professionals break out of their traditional mode of behaviour, and question many of their ideas, values, and methods, so that the rural poor and their knowledge and priorities can become centre-stage in rural development. This proposal is fleshed out through various themes, analysing past errors, and providing advice for future practical action.

Chambers, Robert
Participatory Workshops: A Sourcebook of 21 Sets of Ideas and Activities
London: Earthscan, 2002, ISBN: 1 85383862 4, 224 pp.
Making participation real requires *participatory* workshops, training, and learning. This source book presents some of the author's vast experience in the form of ideas, activities, and tips – both

serious and light-hearted – on topics such as getting started, seating, forming groups, managing large numbers, analysis, feedback, evaluation, and ending. The book is aimed at all those who try to help others learn and change – from participatory teachers and trainers, through organisers, moderators, and facilitators, staff in training institutes, faculty and teachers in universities, colleges, and schools, to those engaged in management training.

Chambers, Robert, Arnold Pacey, and Lori Ann Thrup (eds.)
Farmer First: Farmer Innovation and Agricultural Research
London: Intermediate Technology Publications, 1989, ISBN: 1 85339 007 0, 240 pp.
Farmer First presents a new approach for agricultural research. Starting with farmers' own innovations, followed by contributions from the agricultural and social sciences, ecology, economics, and geography, it makes the case for a 'farmer first' mode to complement conventional procedures for research and transfer of technology.

Clark, Norman, Pakki Reddy, and Andrew Hall
'Client-driven biotechnology research for poor farmers: a case study from India'
International Journal of Technology Management and Sustainable Development 5(2) (2005 [2006]):125–45.
This article explores an attempt to bring biotechnology more directly within the reach of civil society in general, and resource-poor farmers in particular. The Andhra Pradesh Netherlands Biotechnology Programme (APNLBP) was designed to contribute to poverty alleviation through biotechnologies. Emphasis was put on direct interaction with farmers and related stakeholder groups such as NGOs. The article describes the Programme's inception and evolution, outlines key governance aspects, and sets the analytical discussion within the context of modern 'innovation systems' discourse. This case study shows that the degree of connectivity among different stakeholders affects the flow of information across stakeholder groups, which itself often has a major influence on the degree of technological development. It also highlights the importance of institutions and institutional change in enabling successful innovation to take place.

Clay, Edward J. and Benjamin Bernard Schaffer (eds.)
Room for Manoeuvre: An Exploration of Public Policy in Agricultural and Rural Development
Madison, NJ: Fairleigh Dickinson University Press, 1984, ISBN: 083863243 2, 209 pp.
Contributors to this edited volume investigate and question the conceptual frameworks that underpin policy and project-cycle management procedures used in much development planning and implementation.

Cornwall, Andrea
'Whose voices? Whose choices? Reflections on gender and participatory development'
World Development 31 (8) (2003): 1325–42.
Although participation aims to give everyone who has a stake a voice and a choice, initiatives can become driven by particular gendered interests, leaving the least powerful without voice and with little choice. A gender perspective may help to identify strategies for giving voice, and access to decision making, to those who tend to be marginalised or excluded by mainstream development initiatives. The article explores tensions, contradictions, and complementarities between 'gender-aware' and 'participatory' approaches to development. Cornwall suggests that making a difference may come to depend on challenging embedded assumptions about gender and power, and on making new alliances out of old divisions, in order to build more inclusive, transformatory practice.

Davies, Rick and Jess Dart
The 'Most Significant Change' (MSC) Technique. A Guide to Its Use. Version 1.00
(Funded by CARE International; Oxfam Community Aid Abroad, Australia; Learning to Learn, Government of South Australia; Oxfam, New Zealand; Christian Aid, UK; Exchange, UK; Ibis, Denmark; Mellemfolkeligt Samvirke [MS], Denmark; Lutheran World Relief, USA), 2005, 104 pp., available at http://www.mande.co.uk/docs/MSCGuide.pdf (retrieved 17 April 2008)
This publication is aimed at those who wish to use MSC to help to monitor and evaluate social-change programmes and projects, or simply to learn more about how it can be used. The technique is applicable in many different sectors, including agriculture, but especially in development programmes. It can also be applied in many cultural contexts. MSC has been used by a range of organisations in a wide variety of countries.

Elliott, Janice, Sara Heesterbeek, Carolyn J. Lukesmeyer, and Nikki Slocum
Participatory Methods Toolkit: A Practitioner's Manual
Brussels: King Baudouin Foundation and Flemish Institute for Science and Technology Assessment, 2005, ISBN: 90 5130 506 0, 210 pp., available at http://www.viwta.be/files/30890_ToolkitENGdef.pdf (retrieved 17 April 2008)
This manual is aimed at the novice starting out in the world of participation, as well as at the experienced practitioner. It is meant to be a working tool. Individual methods are available as individual PDF files via www.viwta.be or www.kbs-frb.be.

Estrella, Marisol, with Jutta Blauert, Dindo Campilan, John Gaventa, Julian Gonsalves, Irene Guijt, Deb Johnson, and Roger Ricafort (eds.)
Learning from Change: Issues and Experiences in Participatory Monitoring and Evaluation
London: Intermediate Technology Publications, and Ottowa: International Development Research Centre, 2000, ISBN: 088936895 3, 274 pp.
Learning from Change contains case studies and discussions between practitioners, academics, donors, and policy makers in participatory monitoring and evaluation (PM&E). It explores conceptual, methodological, institutional, and policy issues needed to enrich our understanding and practice of PM&E. The three sections provide a general overview of PM&E, synthesising literature surveys and regional reviews of PM&E practice around the world; case studies illustrating a range of settings and contexts of PM&E application; and the key issues and challenges arising from the case studies and discussions, proposing areas for future research and action.

Folke, Steen and Henrik Nielsen (eds.)
Aid Impact and Poverty Reduction
New York: Palgrave, 2006, ISBN 1 4039 7176 5, 264 pp.
A critical book about the impact of aid, and its ability to reduce poverty in recipient countries, written by 11 development researchers on the basis of the research programme at the Danish Institute for International Studies. *Aid Impact and Poverty Reduction* comprises a review of concepts followed by a section on approaches and dilemmas in aid relations, and a section with case studies of aid impact.

Greeley, Martin
'Food technology and employment: the farm-level post-harvest system in developing countries'
Journal of Agricultural Economics 37 (3) (1986): 333–47.
A quantitative social-benefit cost-analysis which shows how economic analysis that concentrates on economic growth can be extended to investigate issues of poverty, income distribution, and equity.

Griliches, Zvi

'Research costs and social returns: hybrid corn and related innovations'
Journal of Political Economy 66 (5) (1958): 419–31.

A classic quantitative rate-of-return study where cost–benefit analysis is used to explore some of the economic dimensions of research-policy analysis. Interestingly, and perhaps because it was written for a political economy journal, the article describes the historical, political, and economic contexts of the analysis, as well as discussing the strengths and weaknesses of this type of quantitative work which arise from inherent problems of attribution and measurement.

Guijt, Irene

Assessing and Learning for Social Change: A Discussion Paper
Randwijk: Learning by Design and Brighton, UK: Institute of Development Studies, 2007, 59 pp., available at http://www.ids.ac.uk/UserFiles/File/participation_team/publications/ASC_low-res_final_version.pdf (retrieved 17 April 2008)

This source book focuses on process-based monitoring and evaluation, covering concepts, approaches, methods, values and skills, applications, and cases. What is particularly interesting about it is that it offers guidance for a critical methodological approach to articulating participants' own theories of social change – an essential starting point for assessing change and learning what to do better. It offers important breakthrough insights into how to move beyond conventional impact assessments, with wide application beyond the cases examined.

Holland, Jeremy with John Campbell

Methods in Development Research: Combining Qualitative and Quantitative Approaches
London: ITDG, 1999, ISBN: 1 85339572 2, 304 pp.

This book draws together lessons about emerging best practice in combining qualitative and quantitative methods and approaches designed to generate 'numbers' from qualitative/ participatory methods and to monitor and evaluate development processes. By drawing on research in many sectors and countries, the book situates development-research issues within debates about development policy and social research. One of its aims was to help to initiate the process of defining best practice in the use of participatory/ qualitative and quantitative methods, and issues of methodological triangulation.

Involve

People & Participation: How to Put People at the Heart of Decision-making
London: Involve, 2005, 114 pp., available at http://www.involve.org.uk/index.cfm?fuseaction=main.viewSection&intSectionID=400 (retrieved 16 April 2008).

This publication is about democratic reform. It provides practical detail, drawing on the experiences of hundreds of practitioners who have used new methods to involve the public in issues ranging from local planning to nano-technology. Its starting premise is that deepening and strengthening democracy depends on success in learning lessons about why some kinds of participation lead to better and more legitimate decisions, while others do not. Greater public involvement can be a great help in addressing some of our most pressing problems, and in countering the risks of distrust and alienation. However, much participation today is superficial: exercises in 'ticking boxes' rather than good democratic governance, or using public consultation to justify decisions that have already been made. *People and Participation* is also available as a website, www.peopleandparticipation.net (see below).

Kabeer, Naila

'Gender equality and women's empowerment: a critical analysis of the third Millennium Goal'
Gender and Development 13 (1) (2005): 13–24, available at http://www.oxfam.org.uk/what_we_do/resources/downloads/gmd-3.pdf (retrieved 17 April 2008)

Kabeer discusses the third Millennium Development Goal (MDG) on gender equality and women's empowerment. She explores the concept of women's empowerment and highlights ways in which the indicators associated with this Goal – relating to education, employment, and political participation – can contribute to it.

Kanji, Nazneen

'Corporate responsibility and women's employment: the case of cashew nuts'
Gender and Development 12 (2) (2004): 82–87.

Falling international prices and buyers' and retailers' exploitative practices have depressed the wages and working conditions of workers in developing countries. Kanji presents and discusses a rare example of better practice in the cashew-nut industry in Mozambique. She demonstrates that collaboration between government, companies, and civil-society organisations at the national level can contribute to gender equality and sustainable development. However, if respectable wages and working conditions are to be provided in a liberalised, market-oriented environment, potentials and constraints need to be analysed across the entire value chain to inform business in developing countries. The main challenge is to strengthen the business incentives for more responsible practice at all levels.

Kanji, Nazneen and Laura Greenwood

Participatory Approaches to Research and Development in IIED: Learning from Experience
London: IIED, 2001, ISBN: 1 899825 81 9, 68 pp., available at http://www.iied.org/pubs/pdfs/9095IIED.pdf (retrieved 17 April 2008)

IIED conducted an internal learning process, examining participatory approaches and methods in 12 research projects involving many IIED staff. This review, primarily intended for internal organisational learning, provides lessons for other organisations which use participatory approaches and methods. Its findings and recommendations cover the meaning and use of the term 'participation'; the requisites for a co-learning approach to collaborative research; the importance of partnerships for positive research outcomes; the factors that support methodological innovation and reflection; trade-offs in the use of participatory methods and approaches; the importance of information and communication; and the factors that constrain the promotion of learning within an organisation.

Laws, Sophie, Caroline Harper, and Rachel Marcus

Research for Development: A Practical Guide
London: SAGE and Save the Children (UK), 2003, ISBN: 07619732 7 3, 488 pp.

This book provides both a quick reference manual and a learning tool for all those engaged in development work at all levels, with emphasis on *practical*. It reviews the complete research process, from outlining the essential role and purpose of research, highlighting issues specific to development research, to demonstrating how to evaluate and draw the best results from research. It includes an overview of different types of research in development work; practical steps in writing a brief and managing research; practical steps in evaluating and promoting research findings; step-by-step guides to getting started and choosing a research method; detailed guidelines to seven key research techniques; examples, exercises, summaries, and checklists; and a glossary and guides to additional resources and packages, including websites dealing with development research.

Lawrence, Anna

'"No personal motive?" Volunteers, biodiversity and the false dichotomies of participation'
Ethics, Place and Environment 9 (3) (2006): 279–98.

This study of voluntary biological monitoring experiences and outcomes finds that they cannot be fitted into the usual dichotomy of 'instrumental' *versus* 'transformative' approaches. Rather, participation can enhance the information base for environmental management; change participants through education about scientific practice and ecological change; lead to changes in life direction or group organisation; and influence decision makers. Personal transformation can take place within a conventional top–down context, while grassroots data collection can reinforce the *status quo* and protect local interests. Partnerships between actors can provide distinct but complementary and mutually rewarding outcomes. Power is not located in a data-consuming centre, and data are not meaningless materials that leave the collector unmoved. A more dynamic model of human-nature relations is presented: one which connects humans and information in the participatory process.

Lewis, David and David Mosse (eds.)

Development Brokers and Translators: The Ethnography of Aid and Agencies
Bloomfield, CT: Kumarian Press, 2006, ISBN: 156549217 X, 251 pp.

Ethnography can be an indispensable tool for understanding the complex and dynamic ways in which communities relate to ideas and resources. The ethnography of development reveals a world of hybrid interests and practices, with no clear division between developers and the developed, perpetrators and victims, domination and resistance, or the incompatible rationalities of scientific and indigenous knowledge, where the realms of reason and the real world merge into each other, and in which rational policy representations often hide the messiness of practice that precedes the ideas and technologies of development.

Mosse, David

Cultivating Development: An Ethnography of Aid Policy and Practice
London: Pluto Press, 2005, ISBN: 074531798 7, 288 pp.

This is an ethnographic study of a project funded by the UK Department for International Development that was seen as a flagship of participatory approaches to agricultural and rural development. The study demonstrates well the complexities of 'evaluations and assessments' and shows why anthropological/ethnographic studies are necessary, not only for understanding processes of social change, but also for understanding and interpreting other forms of investigation such as quantitative economic and natural-resources-based investigations.

Mosse, David, John Farrington, and Alan Rew (eds.)

Development as a Process: Concepts and Methods for Working with Complexity
London: ODI, and New York: Routledge, 1998, ISBN: 041518605 6, 208 pp.

This book covers theoretical and practical issues for monitoring activities, evaluations, assessments, and the like, where interest is in understanding processes of change which lead to achieving development goals.

Neef, Andreas (ed.)

Participatory Approaches for Sustainable Land Use in Southeast Asia
Bangkok, Thailand: White Lotus, 2005, ISBN: 974 4800 67 4, 412 pp.

This book comprises in-depth analyses and discussion of Participatory Research and Development in action, with emphasis on the needs of rural communities in marginal regions of Cambodia, Indonesia, Lao PDR, Nepal, The Philippines, south China, Thailand,

and Vietnam. Thirty-eight scientists and development practitioners share their extensive multi-disciplinary experience and discuss the relevance, application, and pitfalls of participatory approaches *vis à vis* research and development.

Neef, Andreas, Franz Heidhues, Karl Stahr, David Thomas, and Pittaya Sruamsiri (eds.)
Special Issue 'Integrated and Participatory Research Approaches towards Sustainable Livelihoods and Ecosystems in Mountainous Regions', *Journal of Mountain Science* 3 (4) (2006): 276–346.
Chengdu, China: Institute of Mountain Hazards and Environment, CAS, ISSN: 1672-6316.
This special issue brings together a selection of papers presented at the International Symposium 'Towards Sustainable Livelihoods and Ecosystems in Mountainous Regions', held in Chiang Mai, Thailand, 7–9 March 2006. The contributions cover case studies from Europe, Africa, Latin America, South Asia, and Southeast Asia. They all employed various types and levels of integration and participation in the research process, thus providing excellent examples of the potential and challenges of integrated and participatory approaches towards sustainable ecosystems and livelihoods in mountainous regions.

Norton, Andy, Bella Bird, and Karen Brock
A Rough Guide to PPAs – Participatory Poverty Assessment: An Introduction to Theory and Practice
London: Overseas Development Institute (ODI), 2001, ISBN: 085003520 1, 85 pp., available at http://info.worldbank.org/etools/docs/library/238411/ppa.pdf (retrieved 17 April 2008)
This handbook is designed to provide practical guidance for development practitioners. It summarises key messages from recent experience; gives guidance on deciding, at country level, whether a PPA might make a useful contribution to improving the effectiveness of poverty-reduction policy; offers guidance for designing the process to ensure that the PPA will have a beneficial impact on policy; and provides help in finding useful literature and technical assistance. A generic book like this should be more useful early in the process of considering and designing a PPA, rather than later.

Okali, Christine, James Sumberg, and John Farrington
Farmer Participatory Research: Rhetoric and Reality
London: Intermediate Technology Publications, 1994, ISBN: 978185339252 8, 168 pp.
The authors describe and assess how farmer participatory research is presently being used within a broad range of agricultural research and development programmes. They argue for the linkage of project objectives with implementation strategies.

Painter, Genevieve Renard
'Linking women's human rights and the MDGs: an agenda for 2005 from the UK Gender and Development Network'
Gender and Development 13 (1) (2005): 79–93, available at http://www.oxfam.org.uk/download/?download=http://www.oxfam.org.uk/what_we_do/resources/downloads/gmd-9.pdf (retrieved 17 April 2008)
Activists for women's human rights need to recognise that the Millennium Development Goals (MDGs) are a potentially powerful tool for progress on development and human rights, and they need to build on the political will mobilised around them. However, the MDGs reflect the dominant development approach in seeking to use women in their existing social roles to 'deliver' other aims; they do not address the need to eradicate gender inequality, with consequent lack of commitment to address key issues for women, including gender-based violence. There are

also gender-related problems with the MDGs' indicators, analytical approach, and accountability mechanisms. The author seeks a reframing of the MDGs as human-rights obligations through fostering links between the 2005 reviews of implementation of the Beijing Platform for Action and progress on the Millennium Declaration and the MDGs.

Robb, Caroline M. and Miedzynarodowy Fundusz Walutowy
Can the Poor Influence Policy? Participatory Poverty Assessments in the Developing World, 2nd ed.
Washington, DC: The World Bank, 2002, ISBN: 082135000 5, 195 pp.
This book shows how to include the poor by using the Participatory Poverty Assessment (PPA) method. PPA was developed by the World Bank in partnership with NGOs, governments, and academic institutions, and has been implemented in over 60 countries worldwide. The second edition draws on 'new' PPA case examples.

Rodwin, Lloyd and Donald A. Schön (eds.)
Rethinking the Development Experience: Essays Provoked by the Work of Albert O. Hirschman
Washington, DC: The Brookings Institute, 1994, ISBN: 081577551 2, 386 pp.
Provides empirically based information to support many of the insights and understanding about development interventions from the classic work in the 1970s of Albert Hirschman. Many of these are as relevant today as earlier.

SIDA (ed.)
'Discussing Women's Empowerment: Theory and Practice', *SIDA Studies No. 3*
Stockholm: SIDA, 2001, ISBN: 91 586 8957 5, 130 pp., available via http://www.eldis.org/go/display/?id=11046&type=Document (retrieved 17 April 2008)
This volume offers a selection of papers presented at the SIDA conference on 'Power, Resources and Culture in a Gender Perspective: Towards Dialogue Between Gender Research and Development Practice'. Naila Kabeer's overview is an important contribution to the conceptualisation of empowerment.

Van Mele, Paul
'Zooming-in zooming-out: a novel method to scale up local innovations and sustainable technologies'
International Journal of Agricultural Sustainability 4(2) (2006): 131–142, available at http://www.warda.org/warda/IJAS%20zooming-in%20zooming-out.pdf (retrieved 17 April 2008)
The role of video in scaling up sustainable rice technologies is assessed through two case studies from Bangladesh and Benin. Both process and outcomes of participatory research increased the effectiveness of educational videos. The video-production process itself enabled researchers and development workers to learn about local innovations and caused them to change their attitudes towards working with farmers. For increased impact, learning topics should be regionally relevant. The 'zooming-in zooming-out' method starts with a broad stakeholder consultation to define regional issues, after which communities are approached to get a better understanding of their ideas, innovations, and the words they use in relation to the chosen topic (zooming-in). Key learning needs are defined and videos produced in close consultation with the end-users. Showing the draft videos to more villages (zooming-out) generated further insights on the innovations and their socio-cultural context, such that further adjustments could be made. The videos successfully explained underlying biological and physical principles, using a few well-selected local innovations, merged with scientific knowledge. The more the portrayed principles resonated with farmers' existing knowledge and practice, the more useful video

became as a stand-alone method for scaling up. Facilitation increased adoption of sustainable technologies, but was not always a prerequisite.

Van Mele, Paul, Ahmad Salahuddin, and Noel Magor (eds.)
Innovations in Rural Extension: Case Studies from Bangladesh
Wallingford: CABI Publishing, 2005, ISBN: 085119028 2, 307 pp.
'*Innovations in Rural Extension* stands on its head the old linear or pipeline paradigm in which research innovates and passes on innovations to extension for promotion and spread. In the place of these old mindsets and methods are a range of practices and approaches which emphasise listening, learning, negotiating and facilitation, and training of facilitators. The first major section, on gender, gives long-overdue prominence to women in South Asian agriculture. This is followed by accounts of experiences and comparisons between the family approach in extension, participatory video for women-to-women extension, "going public" (seizing the opportunities of public places like markets when many are present), picture songs, village soil fertility maps, the "discovery learning" of farmer field schools, and enterprise webs for analysis of and action on complex relationships. Innovations involved include integrated rice–duck farming, and various aspects of seed systems: building a rice-seed network, a value-chain approach for aromatic rice, and much more. Paradigmatically, *Innovations in Rural Extension* has opened up as never before the need and potentials for methodological pluralism. It shows a wide range of complementary choices of what to do, and it compares their costs and effectiveness. For too long, agricultural extension has been in the doldrums, and agriculture a diminished priority among aid agencies. The editors confront and discuss the issues of extension, complexity and poverty, of creativity and flexibility, and of motivation. Perhaps the most important section, which could have the biggest impact, concerns donors and flexibility, and projects, service providers and potential champions. Lessons and warnings are laid out. The relevant sections, indeed the whole book, should be required reading for all who fund agricultural research and extension.' (Review by Robert Chambers, originally published in *Rice Today* (April–June 2007): 36. Reproduced with permission from the International Rice Research Institute.)

Journals

Gender and Development (http://www.oxfam.org.uk/gadjournal/)
Published three times a year on behalf of Oxfam GB, *Gender and Development* aims to support development policy and practice to further the goal of equality between women and men. Each topic-based issue offers an overview and resources section to complement the main content.

LEISA Magazine (http://www.leisa.info/)
ILEIA, the Centre for Information on Low External Input and Sustainable Agriculture, promotes exchange of information for small-scale farmers in the South through identifying technologies that build on local knowledge and traditional technologies, and involve farmers in development. *LEISA Magazine* offers an opportunity for those working in agricultural development, particularly in diverse, risk-prone, and resource-poor regions, to publish their experiences and to read about those of others. The thematic global edition of *LEISA Magazine* is published quarterly. Regional editions are published together with ILEIA's partner organisations in the South.

Participatory Learning and Action (PLA) Notes (http://www.iied.org/pubs/display. php?o=6098IIED&n=1&l=687&k=PLA%20Notes)
Participatory Learning and Action is a leading informal journal on participatory approaches and methods. It uses expert guest editors for up-to-date accounts of the development and use of participatory methods in specific fields from around the world; and provides a forum for those

engaged in participatory work – practitioners, community workers, activists, and researchers – to share their experiences, conceptual reflections, and methodological innovations.

Websites

ELDIS

http://www.eldis.org/manuals/index.htm

This website aims to 'share the best in development policy, practice and research'. It has an excellent collection of manuals/toolkits for participation.

IDRC Participatory Research Bibliography for Community Based Natural Resource Management

http://www.idrc.ca/fr/ev-3252-201-1-DO_TOPIC.html

This online bibliography provides a list of annotated/abstracted references related to participatory research of potential interest to community-based natural-resource management (CBNRM) researchers. Most annotations are from the author or the publisher; however, some have been written or adapted specifically for the CBNRM research audience.

People and Participation.net

http://www.peopleandparticipation.net/display/Involve/Home

The site provides practical information for those working to involve people. It offers detailed information about a variety of methods as well as case studies. After registering, one can upload case studies, ask questions of experts, and add events. This is currently a 'beta' site, which may change in response to users' feedback.

PovertyNet – Empowerment

http://web.worldbank.org/WBSITE/EXTERNAL/TOPICS/EXTPOVERTY/0,,menuPK:33
6998~pagePK:149018~piPK:149093~theSitePK:336992,00.html

This World Bank portal provides access to a large number of case studies, workshop reports, and methodological tools, including the *Empowerment and Poverty Reduction Sourcebook*, with an emphasis on economic indicators and quantitative methods.

The World Bank: 'Participation and Civic Engagement'

http://web.worldbank.org/WBSITE/EXTERNAL/TOPICS/EXTSOCIALDEVELOPMENT/
EXTPCENG/0,,menuPK:410312~pagePK:149018~piPK:149093~theSitePK:410306,00.html

The Participation and Civic Engagement Group of the Social Development Department promotes the participation of people and their organisations to influence institutions, policies, and processes for equitable and sustainable development. The Group supports World Bank units, client governments, and civil-society organisations to incorporate participatory approaches in the design, implementation, monitoring, and evaluation of Bank-supported operations. The Group works to enhance capacity for participatory processes and social accountability, and develops analytical instruments to assess constraints on the effectiveness of civil society, focusing on the following themes: Social Accountability, promoting the participation of citizens and communities in exacting accountability; Enabling Environment for Civic Engagement, promoting conditions that enable civil society to engage effectively in development policies and projects; Participatory Monitoring and Evaluation, promoting the participation of local beneficiaries in the monitoring and evaluation of projects and programmes; and Participation at the Project, Program, and Policy Level, promoting participatory processes and stakeholder engagement at the project, programme, and policy levels.

Index

Development in Practice Books
Series Editor: Deborah Eade

Each title in the *Development in Practice Books* series offers a focused overview of practice-relevant analysis, experience, and research on key topics in development.

Participatory Research and Gender Analysis
New Approaches
Edited by Nina Lilja, John Dixon, and Deborah Eade

Achieving Education for All through public–private partnerships?
Non-state provision of education in developing countries
Edited by Pauline Rose

10 0700890 2

UNIVERSITY OF NOTTINGHAM

WITHDRAWN

FROM THE LIBRARY

Participatory Research and Gender Analysis

Agricultural development research aims to generate new knowledge or to retrieve and apply existing forms of knowledge in ways that can be used to improve the welfare of people who are living in poverty or are otherwise excluded, for instance by gender-based discrimination. Its effective application therefore requires ongoing dialogue with and the strong engagement of men and women from poor marginal farming communities.

This edited volume discusses opportunities afforded by effective knowledge pathways linking researchers and farmers, underpinned by participatory research and gender analysis. It sets out practices and debates in gender-sensitive participatory research and technology development, concentrating on the empirical issues of implementation, impact assessment, and institutionalisation of approaches for the wider development and research community. It includes six full-length chapters and eight brief practical notes as well as an annotated resources list of relevant publications, organisations, and websites adding to the portfolio of approaches and tools discussed. Most of the 33 contributors work in the specialised agencies that form part of the Consultative Group on International Agricultural Research (CGIAR).

This book is based on a special issue of *Development in Practice,* Volume 18, Numbers 4 & 5 (August 2008).

Nina Lilja is Director of International Agricultural Programs in the College of Agriculture K-State Research and Extension, Kansas State University. She was previously Impact Assessment Economist at the CGIAR Systemwide Program on Participatory Research and Gender Analysis for Technology Development and Institutional Innovation (PRGA Program) in Colombia.

John Dixon is Senior Advisor for the Cropping Systems and Economics (CSE) program and Regional Coordinator, South Asia, at the Australian Centre for International Agricultural Research (ACIAR). He was previously Director of Impacts Targeting and Assessment at the International Wheat and Maize Improvement Center (CIMMYT) in Mexico.

Deborah Eade is a freelance writer and editor on international development and humanitarian issues, based near Geneva. She was Editor-in-Chief of *Development in Practice* from 1991 to 2010, prior to which she spent 10 years working for various NGOs in Mexico and Central America.